Biodiesel: Advanced Approaches and Applications

Biodiesel: Advanced Approaches and Applications

Edited by **Kurt Marcel**

New Jersey

Published by Clanrye International,
55 Van Reypen Street,
Jersey City, NJ 07306, USA
www.clanryeinternational.com

Biodiesel: Advanced Approaches and Applications
Edited by Kurt Marcel

International Standard Book Number: 978-1-63240-080-2 (Hardback)

Printed in the United States of America.

Contents

Permissions

List of Contributors

Preface

This book covers nearly every aspect of biodiesel production and its applications. All facets from the applications of the biodiesel and its by-products are described in the book. This book would appeal to the students, scientists, researchers and professionals who are involved in this genre of science.

The researches compiled throughout the book are authentic and of high quality, combining several disciplines and from very diverse regions from around the world. Drawing on the contributions of many researchers from diverse countries, the book's objective is to provide the readers with the latest achievements in the area of research. This book will surely be a source of knowledge to all interested and researching the field.

In the end, I would like to express my deep sense of gratitude to all the authors for meeting the set deadlines in completing and submitting their research chapters. I would also like to thank the publisher for the support offered to us throughout the course of the book. Finally, I extend my sincere thanks to my family for being a constant source of inspiration and encouragement.

Editor

By-Products Applications

Biomethanol from Glycerol

Joost G. van Bennekom,
Robertus H. Venderbosch and Hero J. Heeres

Additional information is available at the end of the chapter

1. Introduction

Methanol is an important bulk chemical in the chemical industry. The global methanol demand was approximately 32 million metric tons in 2004 and is expected to grow [1]. Methanol is used mainly for the production of formaldehyde, acetic acid, and application products including polymers and paints. Furthermore, methanol can be used as a clean and renewable energy carrier [1]. Methanol is mainly produced from syngas, a mixture of H_2, CO, and minor quantities of CO_2 and CH_4. Syngas is commonly produced from fossil resources like natural gas or coal. Biomass, however, can also be used as resource for syngas and allows the synthesis of green methanol. Green methanol not only has environmental benefits, but may also lead to considerable variable cost reductions if the biomass resource has a low or even negative value.

1.1. Renewable methanol

Methanol synthesis from biomass was already proposed during the first oil crisis in the 1970s [1]. In the 1980s a comprehensive review was published on the production of methanol from syngas derived from wood. Different gasification technologies were proposed and demonstration projects of these technologies were discussed [2, 3]. In the mid 1990s several projects on methanol synthesis from biomass were initiated such as the Hynol project in the USA and the BLGMF (black liquor gasification with motor fuels production) process in Sweden [4-6]. Schwarze Pumpe, Germany developed a process to convert coal and waste, including sewage sludge, to methanol (capacity ± 150 ML/y) [7]. Unfortunately no experimental data of these processes are available in open literature.

Several initiatives were started in the 2000s. At a scale of 4 t/d, Chemrec in Sweden produces methanol and dimethyl ether (DME) since 2011. Syngas is obtained by entrained flow gasification of black liquor [8]. The production of methanol from glycerol is demonstrated on industrial scale by BioMCN in The Netherlands [8]. At BioMCN, the natural gas reforming unit has been modified to enable steam reforming of glycerol. The syngas is converted to methanol in their conventional packed bed methanol synthesis reactors, with a capacity for methanol production of 250 ML/y [8].

The amount of published experimental results on the production of methanol from biomass is rather limited. Most publications on methanol production from biomass are desk-top studies and data comparison is difficult [5, 6, 9-13]. These studies often combine biomass gasification and conventional methanol synthesis with, in some cases, electricity production [5, 6, 11, 13]. Xu et al. conducted an experimental study and demonstrated methanol production from biomass by reforming pyrolysis liquids into H_2 and CO_2 followed by catalytic syngas conditioning to convert part of the CO_2 into CO [14]. Methanol synthesis was conducted in a packed bed reactor, with an overall carbon conversion of around 23% (corresponding with a methanol production rate of 1.3 kg methanol/kg catalyst/h).

An interesting concept of using biomass to produce methanol is the co-processing of biomass and fossil resources, e.g. co-gasification of biomass with coal or natural gas [9, 10, 12]. The advantage of co-feeding natural gas is that the syngas derived will become more suitable for methanol synthesis as syngas from biomass is deficient in H_2 and syngas from natural gas in CO or CO_2.

The concepts involved in the current processes for the synthesis of methanol from biomass generally involve an initial gasification step at elevated temperatures and pressures. The approach demonstrated in this chapter is syngas production through a hydrothermal process, viz. conversion of a wet biomass stream to syngas by reforming in supercritical water (RSCW), followed by high pressure methanol synthesis.

An interesting wet biomass resource is glycerol, the by-product from the biodiesel industry. In Europe, the share of transportation fuel to be derived from renewable resources in 2020 is targeted at 10% [15]. It is expected that biodiesel and ethanol will make up the largest share and consequently Europe's biodiesel production increased significantly in the 2000s (see Figure 1) [16]. In the last few years, though, economics of biodiesel production deteriorated as the income from the sales of glycerol decreased, while the costs of feedstock increased.

As for the production of every (metric) ton of biodiesel, roughly 100 kg of methanol is required and a similar quantity of glycerol is produced, both methanol demand and glycerol production increased. An interesting option addressing the surplus of glycerol and the demand for methanol is to produce methanol from the glycerol. If this process is conducted by the biodiesel producer he will become less dependent on the methanol spot price, there is a (partial) security of supply of methanol, and by-products can be used as a green and sustainable feed product. However, the scale of traditional methanol synthesis (> 2000 t/d) is much larger than the scale of methanol synthesis explored in the Supermethanol project.

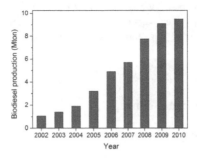

Figure 1. Biodiesel production in Europe. Adapted from reference [16].

The Supermethanol initiative focuses on a small/medium scale biodiesel plant (30,000 – 100,000 t/y) and the aim of the project is to develop a methanol synthesis process using glycerol as feed at a capacity matching the biodiesel production. The glycerol intake for the production of syngas will be in the range of 3000 up to 10,000 t/y [17]. The scope of the Supermethanol concept is schematically outlined in Figure 2.

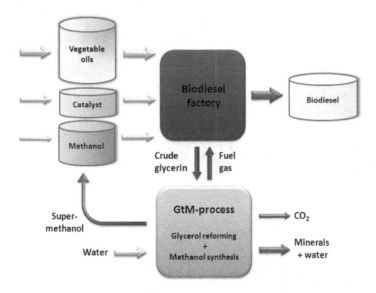

Figure 2. Outline of the Supermethanol concept. The glycerol-to-methanol (GtM-) process is the process under investigation in the Supermethanol project.

The biodiesel factory is the core of the process in which vegetable oils react with methanol in the presence of a catalyst to produce biodiesel and by-product glycerol. The glycerol can be converted into methanol in the glycerol-to-methanol (GtM-) process. This process is an inte gration of two separate processes, viz. the reforming in supercritical water (RSCW) of glycerol to syngas, followed by the conversion of this syngas into methanol. Additional fuel gas is produced, which can be used to generate heat for the biodiesel production or in the GtM-process. A more detailed overview on the GtM-concept is given in the next section.

2. The glycerol-to-methanol concept

2.1. Theoretical considerations

The most attractive syngas for methanol synthesis has a stoichiometric number (S_N), defined in Eq. 1, of approximately 2, which corresponds to the stoichiometric ratio for methanol synthesis.

$$S_N = \frac{\left(H_2 - CO_2\right)}{\left(CO + CO_2\right)} = 2 \tag{1}$$

When glycerol decomposes solely into H_2, CO, and CO_2 the maximum S_N is 1.33. This is illustrated by Eqs. 2 and 3. In Eq. 2, glycerol decomposition into syngas including the reversible water-gas shift (WGS) reaction is given. The syngas composition at equilibrium (neglecting methanation), expressed in terms of x, is a function of the temperature and the water concentration.

$$C_3H_8O_3 \xrightarrow{\ xH_2O\ } (3\text{-}x)\ CO + \left(4\text{+}x\right) H_2 + x\ CO_2 \tag{2}$$

Application of the definition for S_N and introduction of gas phase compositions in terms of x (see Eq. 3) confirms that the S_N value is 1.33 at most and independent of the progress of the WGS reaction:

$$S_N = \frac{\left(H_2 - CO_2\right)}{\left(CO + CO_2\right)} = \frac{\left(4 + x - x\right)}{\left(3 - x + x\right)} = \frac{4}{3} \tag{3}$$

The S_N value, though, can be increased by the addition of H_2 to, or removal of CO_2 from, the syngas. To obtain the highest methanol yield per kg glycerol, glycerol reforming followed by syngas conversion should proceed, for example, according to Eq. 4, where glycerol is selectively converted into H_2, CO, and CO_2. Subsequently, all H_2 and CO react to methanol, while the CO_2 remains.

$$3\ C_3H_8O_3 + 2\ H_2O \rightarrow 7\ CO + 14\ H_2 + 2\ CO_2 \rightarrow 7\ CH_3OH + 2\ CO_2 \tag{4}$$

As a theoretical maximum, 2.33 mol carbon/mol glycerol end up in methanol (77.8% on carbon basis or 0.81 kg methanol/kg glycerol on weight basis). Actual yields, however, will be lower as both processes, glycerol reforming and methanol synthesis, involve equilibrium reactions and the occurrence of other reactions like the formation of higher hydrocarbons, higher alcohols (HA), and methanation.

2.2. Description of the continuous integrated GtM-bench scale unit

The integrated unit consists of a reformer section and a methanol synthesis section. A schematic flow sheet of the unit is given in Figure 3. An extensive description of the reformer section is published elsewhere [18]. The reformer section was operated in continuous mode with a throughput of 1 L aqueous feed/h. Glycerol and water were introduced to the system from feed containers F1 or F2 through a pump and subsequently reformed in five reforming reactors (R1 – R5) in series. The temperature in each reactor can be adjusted individually.

During operation *in situ* separation of the water and gas phase after the reformer section was performed in a high pressure separator (HPS). The liquid phase in the HPS, can either be depressurized and transferred to a low pressure separator (LPS) or recycled via a recycle pump. In the former operating mode (using the LPS), the gases dissolved in the aqueous phase are released, quantified (Gallus G1.6 gas meter), and analyzed (gas chromatography, GC). In the latter operation mode (recycle mode) the gases remain dissolved and fresh glycerol feed is injected in the recycle stream before the first reforming reactor (R1). If required all reforming reactors can easily be filled with catalyst. The gas phase from the HPS was directly fed to the methanol synthesis section without upgrading or selective removal of components.

The methanol section contains three tubular packed bed reactors (P1 – P3, each $L = 500$ mm, $ID = 10$ mm) surrounded by heating jackets. A heating/cooling medium was flown through the heating jackets to control the temperature in the reactors. Temperatures were recorded at 4 positions inside packed bed P2 (at locations 2 to 30 cm from the entrance) and at the exit of packed bed reactors P2 and P3. Two or three of the tubular reactors were filled with catalyst particles ($1 < d_p < 3$ mm). The mixture of methanol, water, and unconverted gases leaving the last packed bed (P3) was cooled (cooler C2) using tap water, depressurized and cooled (cooler C3) to temperatures below 263 K to trap all condensables. Liquid samples were collected in a vessel and the unconverted gas was quantified (Gallus G1.6 gas meter) and analyzed by GC. The methanol synthesis reactors were operated at temperatures of the heating medium between 473 – 523 K and at similar pressure as the reformer section. Several process parameters were logged during operations and the locations, where they were measured, are indicated with letters in bold in Figure 3.

The process pressure was the average of $A_{1,j}$ the temperature of reactor R5 (T_{R5}) was measured at **B** at the end of reactor R5, the glycerol feed flow at **C**, the gas flow of the HPS and LPS at D_1 and D_2 respectively, the temperature at the end of the methanol synthesis bed at **E**, the amount of liquid product at **F**, and the unconverted gas flow at **G**.

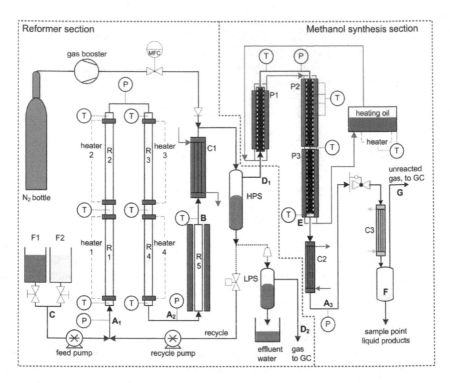

Figure 3. Flow sheet of the integrated GtM-bench scale unit. HPS and LPS refer to high pressure separator and low pressure separator respectively. F = feed container, C = cooler, P = packed bed reactor, R = reforming reactor. The bold capital letters correspond to the locations where relevant process parameters were measured [8].

2.3. Analyses

The composition of the off-gas from the reforming section and methanol synthesis section was analyzed using an online dual-column gas chromatograph (GC 955, Syntech Spectras) equipped with thermal conductivity detectors. CO was analyzed and quantified using a molsieve 5 Å column (L = 1.6 m) with helium as carrier gas. CH_4, CO_2, and C_{2+} were analyzed on a Chromosorb 102 column (L = 1.6 m) with helium as carrier gas. H_2 was analyzed on the molsieve column using argon as carrier gas. The total organic carbon (TOC) content of the effluent water from the RSCW process was analyzed using a TOC analyzer (TOC-V_{CSN}, Shimadzu). The water content of the methanol was determined by Karl Fischer-titration. The composition of the organics in the liquid phase after the methanol synthesis reactor was determined with a GC (HP 5890 series II) equipped with a flame ionization detector (FID) over a Restek RTX-1701 column (L = 60 m, ID = 0.25 mm) coupled with a mass spectrometer (MS, HP 5972 series). The FID was used for the quantification of the components and the MS for the identification of the components. The FID was calibrated for the main constituents of the organic fraction: methanol, ethanol, 1-propanol, 2-propanol, 1-butanol, 2-butanol, 1-pentanol, 2-methyl-1-propanol, 2-methyl-1-butanol.

2.4. Definitions

The carbon conversion of glycerol (ζ_{gly}) in the reformer section is defined as the difference between the molar carbon flow of glycerol in the feed and the effluent ($\phi_{C,gly} - \phi_{C,effluent}$) over the molar carbon flow of glycerol in the feed ($\phi_{C,gly}$):

$$\zeta_{gly} = \frac{\phi_{C,gly} - \phi_{C,effluent}}{\phi_{C,gly}} \cdot 100\% \tag{5}$$

The overall conversion of carbon in glycerol to carbon in methanol (ζ_C) in the integrated unit is the molar carbon flow in methanol ($\phi_{c,MeOH}$) over the molar carbon flow of glycerol in the feed.

$$\zeta_C = \frac{\phi_{C,MeOH}}{\phi_{C,gly}} \cdot 100\% \tag{6}$$

The methanol yield (η) is the mass flow of methanol (ϕ_{MeOH}) produced over the mass flow of glycerol (ϕ_{gly}) fed.

$$\eta = \frac{\phi_{MeOH}}{\phi_{gly}} \tag{7}$$

The conversion of gas component i (ζ_i) in methanol synthesis is defined as the molar conversion rate ($\phi_{i,in} - \phi_{i,off}$) over the molar flow of component i originally present ($\phi_{i,in}$) after glycerol reforming.

$$\zeta_i = \frac{\phi_{i,in} - \phi_{i,off}}{\phi_{i,in}} \cdot 100\% \tag{8}$$

The carbon selectivity towards product i (σ_i) is defined as the molar carbon flow of product i ($\phi_{C,i,off}$) over the molar carbon flow of glycerol in the feed.

$$\sigma_i = \frac{\phi_{C,i,off}}{\phi_{C,gly}} \cdot 100\% \tag{9}$$

2.5. Research strategy

The integration of syngas production in an RSCW-process and syngas conversion in methanol synthesis is the core of the GtM-process. In the RSCW of glycerol a high pressure syngas is produced. The use of this high pressure syngas has distinct advantages for methanol synthesis, which will be dealt with in Section 4. However, before successful integration both processes need to be optimized separately, which was done at the laboratories of the Biomass Technology Group (BTG) in The Netherlands. A unit was available to investigate both process separately before integraton. Results obtained for each process were used to optimize the overall process and maximize the overall carbon conversion (ζ_C), which was the main focus of the research study on the integrated process.

3. Glycerol reforming in supercritical water

3.1. Introduction to reforming

Water becomes supercritical at conditions above its critical temperature (T_c = 647 K) and critical pressure (P_c = 22.1 MPa). In the phase diagram in Figure 4 the square area in the upper right corner represents the supercritical area of water [19].

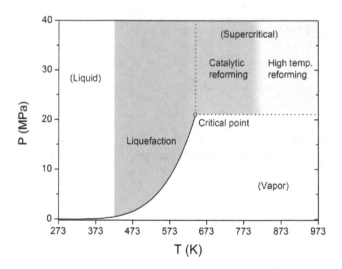

Figure 4. Phase diagram of water. Phases are indicated in parenthesis. Hydrothermal processes with their typical conditions are indicated in the colored areas. Adapted from reference [19].

Typical process conditions for three processes considered for wet biomass valorization are indicated here, viz. liquefaction, catalytic reforming, and high temperature reforming. Hydrothermal liquefaction is conducted at temperatures below and pressures above the critical

point, where biomass is degraded to yield mainly bio-crude (a viscous water-insoluble liquid), char, water-soluble substances, and gas [20]. RSCW of biomass is aimed at gas production and is carried out at conditions beyond the critical point. Here, water acts both as
reaction medium and reactant. RSCW of biomass can be subdivided into catalytic reforming
and noncatalytic reforming. Catalytic reforming is predominantly carried out at the lower
temperature range, while noncatalytic or high temperature reforming is conducted at the
higher temperatures (see Figure 4 [19]).

RSCW is characterized by the occurrence of many reactions, proceeding both in series and in
parallel. The overall reaction of an actual (biomass) feed to liquid and gas phase products is
shown in Eq. 10.

$$C_xH_yO_z + a\,H_2O \rightarrow \alpha\,CO + \beta\,H_2 + \gamma\,CO_2 + \delta\,H_2O + \varepsilon\,CH_4 + \eta\,C_{x'}H_{y'}O_{z'} \tag{10}$$

By-products ($C_{x'}H_{y'}O_{z'}$) are low molecular weight organic compounds, polymerized products, higher hydrocarbons ($x' \geq 2$, $z'=0$), or elemental carbon ($y'=z'=0$). Some of the low molecular weight organics can react further to gas phase components. Subsequent reactions of the
gas phase components may also occur. The following gas phase reactions may occur, depending on process conditions [21]:

$$CO + H_2O \leftrightarrow CO_2 + H_2 \tag{11}$$

$$CO + 3\,H_2 \leftrightarrow CH_4 + H_2O \tag{12}$$

$$CO_2 + 4\,H_2 \leftrightarrow CH_4 + 2\,H_2O \tag{13}$$

$$2\,CO \leftrightarrow CO_2 + C \tag{14}$$

$$CH_4 \leftrightarrow C + 2\,H_2 \tag{15}$$

$$CO + H_2 \leftrightarrow C + H_2O \tag{16}$$

The individual reaction rates depend on operating conditions and the presence of catalysts.
A number of parameters affect the carbon conversion in RSCW, such as feedstock type, feed
concentration, operating conditions, presence of catalysts or catalytic surfaces, and interaction between different components. The state of the art of RSCW in the 2000s has been reviewed extensively in several publications [19, 22-28].

3.2. Reforming of pure glycerol and crude glycerin

The reforming experiments were carried out using only the reforming section of the unit depicted in Figure 3. Typical conditions were temperatures of 723 – 923 K, residence times between 6 – 45 s, and feed concentrations of 3 – 20 wt%. The pressure was around 25 MPa. Two different types of glycerol were used, viz. pure glycerol and crude glycerin. Crude glycerin is glycerol derived from biodiesel production. The crude glycerin used in this research study contains approximately 5 wt% NaCl. The presence of alkali (in this case Na^+) influences mainly the WGS reaction (see Eq. 11). The main gas products for pure glycerol and crude glycerin were: H_2, CO, CO_2, CH_4, and C_2H_6. At complete conversion, roughly 2 mol of carbon in glycerol are converted to carbon oxides while 1 mol of carbon ends up as a hydrocarbon. The gas composition/yield appeared to be a function of the conversion and independent of the feed concentration. The conversion is a function of the process severity, a combination of the residence time and operating temperature. In Figure 5 the gas yields (mol gas/mol glycerol) of the two types of glycerol are depicted. The trend lines shown are fitted to experimental data points [18]. The differences between pure glycerol and crude glycerin can be mainly attributed to the extent of the progress of the WGS reaction [18].

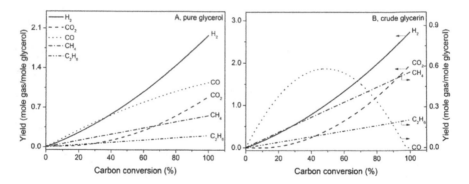

Figure 5. Relations between the conversion and the gas yield for pure glycerol (A) and crude glycerin (B) [29].

From reforming studies with methanol as model compound it was concluded that the hydrocarbons present in the gas mixture in case of glycerol reforming appear to be primary gas phase products. In the methanol reforming experiments in the same unit, gas mixtures with similar H_2, CO, and CO_2 ratios were obtained, but hydrocarbons were hardly observed. Thus, gas phase reactions producing hydrocarbons, e.g. methanation, hardly proceed in the system, which indicates that hydrocarbons are primary gas phase products formed upon glycerol decomposition and not or only to a small extent by gas phase reactions. In glycerol reforming the reactions depicted in Eqs. 10 and 11 proceed, while Eqs. 12 and 13 barely take place. Coke formation was not observed at all and most probably the reactions depicted in Eqs. 14-16 do not proceed.

Based on the experimental results, a simplified reaction scheme for the decomposition of glycerol with a focus on gas production was established and is given in Figure 6. More information on the selection of primary and secondary gas phase products can be found in literature [18].

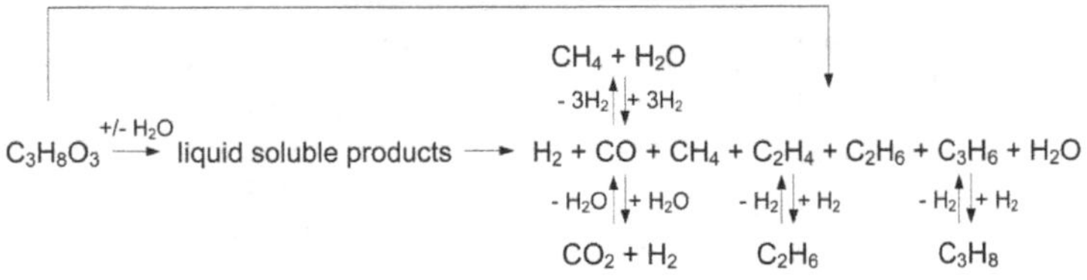

$$C_3H_8O_3 \xrightarrow{+/- H_2O} \text{liquid soluble products} \longrightarrow H_2 + CO + CH_4 + C_2H_4 + C_2H_6 + C_3H_6 + H_2O$$

$$CH_4 + H_2O$$
$$- 3H_2 \uparrow \downarrow + 3H_2$$

$$- H_2O \uparrow \downarrow + H_2O \qquad - H_2 \uparrow \downarrow + H_2 \qquad - H_2 \uparrow \downarrow + H_2$$
$$CO_2 + H_2 \qquad\qquad C_2H_6 \qquad\qquad C_3H_8$$

Figure 6. Decomposition pathways for glycerol in SCW to gaseous products including possible follow-up reactions [18].

In this scheme, CH_4 is shown as a primary product, but can also be formed as a secondary product by methanation. Furthermore, water is produced and the WGS reaction and alkene hydrogenation are included. It is suggested that glycerol can either decompose into liquid soluble products that react further to gas products or that glycerol can directly decompose into gas products. In practice probably both reaction pathways occur. The overall mechanism at complete conversion of glycerol decomposition proceeds through the dehydration of 1 mol H_2O/mol feed [18].

An important quality indicator for the gas composition is the S_N value as defined in Eq. 1. A S_N as close as possible to 2 is desired for methanol synthesis, but as was shown in Eq. 3 a S_N value of 1.33 is the maximum for gas derived in glycerol reforming. The experimental S_N as function of the glycerol conversion is depicted in Figure 7. It can be seen that for both types of glycerol the S_N decreases with increasing conversion. The values are almost equal up to 60%, but differ considerably at the higher conversion. The most attractive S_N's are obtained at the lower conversion although they remain below 1. The S_N can be improved by suppressing the formation of hydrocarbons, which is a challenge as hydrocarbons are formed as primary gas phase products.

3.3. Catalytic reforming

Catalytic reforming was investigated to improve the quality of the syngas obtained in the noncatalytic reforming experiments for subsequent methanol synthesis. An extensive description of catalytic reforming using five different catalysts is given in literature [29]. The catalytic experiments were conducted at temperatures between 648 to 973 K at pressures between 25.5 - 27.0 MPa. The feed concentration was 10 wt%, and the residence time varied from 8 to 87 s. The experiments were conducted using only three reactors (R2, R3, and R4 in Figure 3), with only the latter two reactors containing catalyst. The catalysts clearly promote the glycerol decomposition rate and higher conversion were measured compared to noncatalytic reforming. A typical figure with the gas concentration as a function of the temperature for a Ni based catalyst is given in Figure 8. The equilibrium curves were calculated using a model described in literature [18].

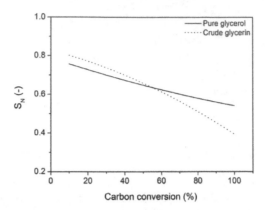

Figure 7. S_N for pure glycerol and crude glycerin as a function of the carbon conversion.

The gas composition is a function of the temperature and in this case equilibria are reached at temperatures exceeding 780 K. This catalyst strongly promotes methanation, as the CH_4 concentration is much higher than in noncatalytic reforming. At the higher temperatures the CH_4 concentration goes down according to thermodynamics. When a Ni based catalyst is used the WGS reaction (Eq. 11) is at equilibrium and almost all CO is converted into CO_2. After the reforming experiments traces of coke were visually observed at the catalyst surface and the reactor wall.

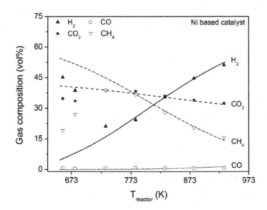

Figure 8. Gas concentration as a function of the temperature for a Ni based catalyst. $P = 25.5 - 27.0$ MPa, [glycerol] = 10 wt%. The curves represent equilibrium compositions, the symbols are experimental points [29].

The performance of this catalyst expressed in S_N, hydrocarbon content, and conversion is given in Figure 9. The conversion is almost complete over the whole temperature range. The concentration of hydrocarbons reaches a maximum as a function of the temperature and decreases at the higher temperatures. The S_N has the inverse profile and increases with higher temperatures, but still remains below 1.

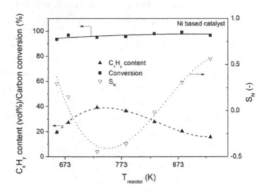

Figure 9. Performance indicators. P = 25.5 – 27.0 MPa, [glycerol] = 10 wt%, ϕ_{gly} = ± 100 g/h. The lines are trend lines and for illustrative purposes only.

3.4. Consequences of the reforming process for the integrated concept

The gas composition in noncatalytic reforming appeared to be not very attractive for methanol synthesis mainly due to the formation of hydrocarbons. Over a Ni based catalyst, CH_4 was the only hydrocarbon present in the gas phase. Its maximum concentration was close to 40 vol% at 730 K. A temperature increase resulted in a decreasing CH_4 concentration. A further decrease can be realized at higher temperatures and lower feed concentrations. Reduction of the hydrocarbon content has a positive effect on the S_N. At feed concentrations around 4 wt% and temperatures of approximately 1000 K, a S_N above 1 can be obtained. These conditions are the most attractive from a gas composition point of view (see also Section 5).

4. Methanol synthesis

4.1. Introduction to methanol synthesis

Methanol synthesis is conducted generally in catalytic gas-solid packed bed reactors. Three equilibrium reactions, taking place at the catalyst surface, are important: (i) the hydrogenation of CO (Eq. 17), (ii) the hydrogenation of CO_2 (Eq. 18), and (iii) the WGS reaction (Eq. 11):

$$CO + 2\,H_2 \leftrightarrow CH_3OH \tag{17}$$

$$CO_2 + 3\,H_2 \leftrightarrow CH_3OH + H_2O \tag{18}$$

All reactions are exothermic. The conversion of $CO+CO_2$ at chemical equilibrium is a function of pressure, temperature, and gas composition (see Figure 10).

Methanol synthesis at industrial scale was initiated by BASF in the 1920s. The operating temperatures were high (573 – 633 K) because of the low catalyst activity [30, 31]. High pressures (15 – 25 MPa) were needed to obtain reasonable conversions. When more active Cu based catalysts and better syngas purification techniques became available, the operating temperature and pressure could be reduced. This development led to the so-called low pressure methanol synthesis process (5 – 10 MPa, 490 – 570 K) which was developed by ICI in the 1960s. Since then, most high pressure units have been converted to low pressure systems [31, 32]. Both synthesis processes require large recycle streams of unconverted syngas due to the limited conversion per reactor pass as is shown in Figure 10 [32]. The reactor temperatures can be lowered further (to 463 – 520 K) due to the recent development of more active catalysts.

In the research study described in this chapter, a combination of low temperatures (468 – 545 K) and high pressures (15 – 25 MPa) is investigated. At this combination of pressure and temperature the Equilibria conversions towards methanol are high (see Figure 10).

4.2. Methanol condensation

The methanol conversion in conventional methanol synthesis is restricted by the chemical equilibrium as shown in Figure 10. There are several opportunities to circumvent the limitations imposed by thermodynamic equilibria and they mainly involve *in situ* removal of methanol. This can be done, for example, by methanol adsorption on fine alumina powder or dissolving methanol in tetraethylene glycol, n-butanol, or n-hexane [33-35]. Another method involves *in situ* condensation at a cooler inside the reactor [36]. With all the different methods mentioned higher syngas conversions were obtained but at the same time all methods have drawbacks including the use of other chemicals, complicated operation procedures, or low yields. Conversions beyond the chemical equilibrium can be obtained with *in situ* condensation of methanol and water without adsorbents or coolers. Condensation occurs at a combination of high operating pressures and low temperatures. Condensation has only been shown indirectly in literature by experimental observations of conversions beyond equilibrium or theoretical models [37-40]. We've demonstrated *in situ* methanol condensation visually in a view cell reactor. In this reactor a propeller-shaped stirrer was equipped with catalyst pellets. The view cell was operated semi-batch wise. Methanol synthesis started when syngas ($H_2/CO/CO_2$ = 70/28/2 vol%) was fed to the reactor. The most striking observation was *in situ* condensation at 20.0 MPa and 473 K (Figure 11). Liquid formation was also observed at 17.5 MPa and 473 K and for other gas compositions [41].

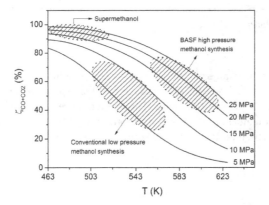

Figure 10. Equilibria in methanol synthesis. Approximate conditions are given for (i) conventional processes, (ii) BASF's high pressure process, and (iii) Supermethanol (this work). Syngas: $H_2/CO/CO_2/CH_4 = 67/24/4/5$ vol% [8].

The liquid accumulated in the view cell upon prolonged reaction times. Part of the catalyst became immersed and even after complete immersion methanol synthesis from syngas bubbling through the liquid went on.

Figure 11. Liquid formation in a view cell. $P = 20.0$ MPa, $T = 473$ K, Syngas: $H_2/CO/CO_2 = 70/28/2$ vol%.

The liquid phase consists of mainly methanol and water. The exact composition is a function of conversion, process conditions, and syngas composition. Condensation may have positive effects on methanol synthesis as will be demonstrated later on in this chapter. Conversions higher than the chemical equilibrium are achieved and almost complete conversion of the limiting component(s) can be obtained at appropriate conditions. As a consequence, recycle and purge streams are not necessary, the limitations on the S_N become less strict, and methanol yields may be increased for a given reactor volume. Most probably the reaction rates in methanol synthesis will be higher at high pressure than at conventional conditions due to higher partial pressures of the reactants, however, experimental validation is required to validate this hypothesis.

4.3. Modelling simultaneous phase and chemical equilibria

A solution model to calculate equilibrium conversions in methanol synthesis including condensation was developed. The effects of process conditions (pressure, temperature, gas composition) can be assessed with the model. Dew points were calculated using Eq. 19 for a given pressure and temperature for each component [42, 43]. A modification of the Soave-Redlich-Kwong equation of state (for polar components) was used to calculate the fugacities of each phase.

$$f_i^V = f_i^L \tag{19}$$

Where, f_i is the fugacity of component i, V and L denote the vapor and the liquid phase respectively. The simultaneous chemical and phase equilibria were calculated using Eq. 19 and theoretical equilibrium constants [44]. The model is described and explained in more detail in the literature [45].

A typical equilibrium diagram for $H_2/CO/CO_2/CH_4$ = 70/5/20/5 vol% is given in Figure 12. The equilibrium diagram for this gas composition illustrates clearly the influence of condensation on the equilibrium conversion.

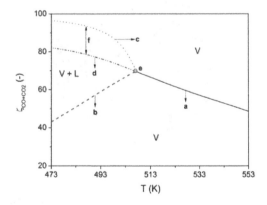

Figure 12. Example of an equilibrium conversion diagram including chemical and phase equilibria. Gas phase equilibrium curve (**a**), dew point curve (**b**), equilibrium curve including liquid formation (**c**), extrapolation of the gas phase equilibrium curve (**d**), point where all equilibrium curves merge (**e**), difference between extrapolated gas phase equilibrium and equilibrium including liquid formation (**f**). P = 20.3 MPa, syngas: $H_2/CO/CO_2/CH_4$ = 70/5/20/5 vol% [41].

In the diagram, 4 curves are shown. Curve **a** (solid curve) is the gas phase equilibrium curve. Curve **b** (dashed curve) is the conversion at which a dew point occurs. Curve **c** (dotted curve) is the equilibrium curve including condensation and curve **d** (dashed-dotted curve) is an extrapolation of the gas phase equilibrium curve. The 4 curves come together in point **e**. In this point ($T = 507$ K), the gas composition is at equilibrium and the dew point temperature of the mixture equals the reactor temperature. Curve **d**, the extrapolation of curve **a**, is the equilibrium conversion when condensation is neglected. When condensation occurs, the equilibrium conversion is much higher which is indicated by the arrow marked **f** (difference between curve **c** and **d**) in Figure 12. The value for **f** amounts to 13.9% at 473 K ($\zeta_{CO+CO2} = 82.6\%$ vs. $\zeta_{CO+CO2} = 96.5\%$) for this particular gas composition.

4.4. High pressure methanol synthesis in a packed bed reactor

Methanol synthesis experiments were conducted in the packed bed reactor using 3 different syngases (see Table 1) and pressures of about 20 MPa. Gas 1 and 2 represent typical methanol synthesis gases ($S_N = 2.0 – 2.3$), with gas 1 rich in CO and gas 2 rich in CO_2. The composition of gas 3 resembles a typical syngas obtained in the reforming of glycerol or biomass in general. For this gas $S_N < 2$ and H_2 is the limiting component. The experiments were performed with a large amount of catalyst to approach the equilibrium conversion.

To check the assumption of equilibrium at the reactor outlet and that the experiments were not conducted in the kinetic regime experiments with different flow rates were conducted for gas 1. For the experiments, methanation and the formation of higher hydrocarbons were negligible. For gas 1, the $CO+CO_2$ conversion at 468 K and 20.7 MPa was 99.5%, which is 7.7% higher than the equilibrium conversion calculated at 7.5 MPa. For gas 2, the difference between methanol synthesis at 20 and 7.5 MPa is more pronounced. The $CO+CO_2$ conversion at 484 K was 92.5% which is 46.9% higher than the equilibrium conversion predicted at 7.5 MPa.

Gas	H_2	CO	CO_2	CH_4	C_2H_6	S_N	Remarks
	(vol%)	(vol%)	(vol%)	(vol%)	(vol%)	(-)	
1	67.0	24.4	3.5	5.1	-	2.3	Industrial gas
2	69.9	5.0	20.0	5.1	-	2.0	CO_2 rich gas
3	54.2	28.9	10.9	4.0	2.0	1.1	Simulated RSCW gas

Table 1. Compositions of the different gases used in methanol synthesis.

The experimental and predicted equilibrium conversions for gas 1 and gas 2 are shown in Figure 13. At temperatures above 495 K for gas 1 and 507 K for gas 2 only a gas phase is present, while at lower temperatures condensation occurs. Methanol production continues in the two phase system until phase equilibrium and chemical equilibrium are reached. At the lowest temperatures in the range, the $CO+CO_2$ conversion is nearly complete for gas 1 (Figure 13A). The conversion of $CO+CO_2$ decreased with increasing temperature as dictated by thermodynamics. The experimental conversions coincide nicely with the conversion predicted by the model. The effect of condensation was more pronounced for gas 2 (Figure

13B). Here, the experimental conversion is even 12.7% higher than the extrapolated gas phase equilibrium curve ($\zeta = 79.8\%$ vs. 92.5% at 484 K).

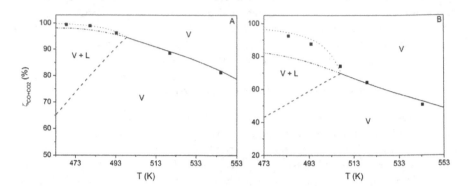

Figure 13. Equilibrium diagrams for methanol synthesis. Gas 1 (A), $P = 19.7$ MPa [41]. Gas 2 (B) [45], $P = 20.3$ MPa. Symbols: experimental data; curves: model results.

In Figure 14, both the $CO+CO_2$ conversion (Figure 14A) and the H_2 conversion (Figure 14B) are depicted for gas 3. At the two lower temperatures, the equilibrium predictions coincide nicely with the experimental data for, but this changes at the higher temperatures most probably due to the formation of HA. This is a common phenomenon for systems at higher temperatures with high CO partial pressures [46, 47]. HA formation is not included in the equilibrium model and this explains the deviations at higher temperatures.

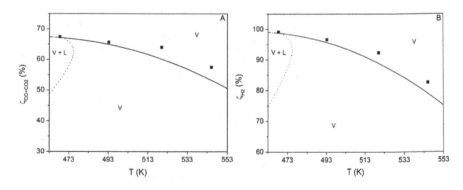

Figure 14. Equilibrium diagram for methanol synthesis from gas 3 including the dew point curve for ζ_{CO+CO2} (A) and ζ_{H2} (B). Symbols: experimental data; curves: model results. $P = 19.4$ MPa [45].

The concentration of the main HA is given in Figure 15. The methanol concentration clearly decreases over the temperature range in favor of the HA. Based on thermodynamics the formation of higher alcohols is expected as HA formation is favored over methanol [48]. The main HA formed in the experiments were ethanol and 1-propanol followed by 1-butanol, 2-methyl-1-propanol, and 2-methyl-1-butanol.

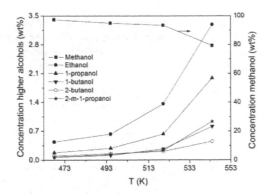

Figure 15. Methanol and HA concentrations as a function of the temperature for gas 3. $P = 19.9$ MPa.

4.5. Consequences of methanol synthesis for the integrated concept

The main result from high pressure methanol synthesis experiments is the observation that high conversions of the limiting component are attainable. These conversions are higher than calculated on the basis of chemical gas phase equilibria and are due to condensation. When the HA concentration was low, the experimental conversion corresponded nicely with the equilibria predicted. In conclusion, when a combination of high pressure and moderate temperature ($463 \leq T \leq 500$ K) is used, high conversions for glycerol derived syngases are expected.

5. Demonstration of the integrated concept

To demonstrate the integrated concept for methanol synthesis, experiments were conducted in the integrated continuous unit depicted in Figure 3 [8]. Both, process conditions and recycle options of the reforming section were investigated to maximize methanol yields. This requires proper operating conditions for each reactor section (reforming, methanol synthesis) to limit by-product formation (e.g. CH_4, higher hydrocarbons, and HA) and to allow operation at high equilibrium conversions in methanol synthesis. In this section, both the results for the overall integrated process will be discussed as well as the result for the reformer section in these experiments. Four different cases are considered with different feed concentrations and operating conditions. An overview of the experiments is presented in Table 2 and a short resume is given below.

The first case is considered the base case. The experiment of the base case was conducted using the LPS without recycling the effluent water (see Figure 3). Part of the gas produced is lost as it dissolves in the aqueous phase that leaves the process. In case 2, 3, and 4 the effluent water from the reformer is recycled after the HPS. Operating with a recycle stream at high pressure in RSCW is a unique feature. As a consequence of the recycle stream no gas is lost in the reformer section through the LPS and additionally, the water consumption of the process is reduced significantly. In these experiments the glycerol reforming is carried out catalytically by using the Ni based catalyst in combination with higher temperatures. All C_{2+} hydrocarbons are then reformed and the CH_4 equilibrium concentration decreases (with higher temperatures) yielding a more attractive gas composition [29].

Finally, in case 4 an extra methanol synthesis packed bed (P1) was added to achieve equilibrium gas phase conditions at the outlet. The methanol synthesis section now consists of 3 packed bed reactors (P1 – P3) in series. The three packed beds were operated at different temperatures, viz. ± 518 K (P1), ± 503 K (P2), and 481 – 482 K (P3) by cooling the heating medium between the reactors. The reaction rate in methanol synthesis depends strongly on temperature. Higher temperatures lead to higher reaction rates. In the first reactor (P1) the reaction rate will be relatively high while the second (P2) and third (P3) are used to achieve equilibrium. Typical run times for the experiments were 6 – 10 h and steady state was reached in approximately 2 h. In case 4 the operating time exceeded 20 h of which 16 h in the integrated mode. This experiment is considered as the long duration experiment.

Case	Catalyst [a] (g)			Recycle	Catalyst [b] (g)		
	R3 [d]	R4 [d]	R5 [d]		P1 [d]	P2 [d]	P3 [d]
1	-	-	-		-	50	51
2	10	10	3	yes	-	50	51
3	10	10	3	yes	-	50	51
4[c]	10	10	3	yes	51	50	51

[a]Ni based catalyst.

[b]A commercial $Cu/ZnO/Al_2O_3$ catalyst.

[c]The catalyst in the reforming section was replaced by fresh catalyst.

[d]R = reformer, P = packed bed reactor for methanol synthesis.

Table 2. Overview of experiments in the integrated unit. R3-5 and P1-3 correspond with the reactors in Figure 3 [8].

5.1. Reformer performance

Typical conditions for the reformer section (see Figure 3) for these experiments were pressures from 24 to 27 MPa and temperatures between 948 and 998 K. At these conditions the residence times of the reformer section (R1 – R5) were in the range of 30 – 35 s. The composition and quantity of the off-gas were analyzed to determine the carbon balance. The hydrocarbon concentration in the off-gas is the summation of the concentrations of CH_4 and C_2H_6. The main results of the glycerol reformer section are summarized in Table 3.

Carbon balance closure for the reformer is very satisfactorily and was between 95 and 104%. The glycerol conversion was almost complete for all experiments, which is in line with previous work [18, 29]. The syngas produced had the following composition range: $H_2/CO/CO_2/C_xH_y = 44 - 67/1 - 21/16 - 34/2 - 18$ vol%, $0.7 \leq S_N \leq 1.2$. The results for each case will be discussed separately in the following section.

5.1.1. Base case

The base case experiment was conducted with a glycerol feed concentration of 10 wt% at 27 MPa and 948 K. The unit was operated without a catalyst in the reformer section and in a once-through mode. Part of the gas (ca. 13%) dissolved in the effluent stream from the LPS (see Figure 3, D_2) and is not used for the subsequent methanol synthesis. Due to the absence of catalysts, a significant amount of CO is present in the syngas. The product gas has a relatively low S_N value of 0.7, which is mainly caused by the formation of hydrocarbons (18 vol % consisting for approximately ⅔ of CH_4 and ⅓ of C_2H_6).

Case	P_a	T_{R5}	ϕ_{gly}	[Gly.]	H_2	CO	CO_2	C_xH_y	S_N	ζ_{gly}	C_{bal}
	(MPa)	(K)	(g/h)	(wt%)			(vol%)		(-)	(%)	(%)
1	27	948	106	10.4	44	21	17	18	0.7	96[b]	96
2	24	998	97	± 10	55	2	32	11	0.7	99.9[c]	97
3	24	998	35	± 4	59	1	34	6	0.8	99.9[c]	104
4	26	998	35	± 4	66	1	30	3	1.2	99.9[c]	95
Loc.[d]	A_{1-3}	B	C		D_1	D_1	D_1	D_1			

[a]The pressure is an average pressure. The actual operating pressure is the indicated pressure ± 1 MPa.

[b]Based on carbon content in the effluent water.

[c]Experiment conducted in recycle mode. Glycerol conversion is estimated based on previous work [29].

[d]Locations where the parameters were measured (see Figure 3).

Table 3. Results of the reforming section (before methanol synthesis) [8].

5.1.2. Case 2 – 4

The intention for case 2 was to aim for higher S_N values. A Ni based catalyst was added to reactor R4 and R5 to reform the higher hydrocarbons. As a consequence the WGS reaction also reached equilibrium and almost all CO was converted into CO_2 [29]. The temperature of reactor R5 was increased with 50 K compared to the base case to aim for a more advantageous equilibrium composition (less CH_4). Furthermore, the effluent water was recycled at high pressure. Recycling the effluent water drastically reduces the water consumption of the process. The recycle flow was adjusted in such a way that the aqueous reactor inlet flow was comparable to the inlet flow in the base case. Compared to the base case no gas was lost through the effluent stream from the LPS. The gas composition obtained over this catalyst differed substantially from the base case. The CO concentration was reduced from 21 vol% to 1 – 2 vol% and the concentration of the hydrocarbons was significantly lower and approached equilibrium ($C_2H_6 \approx 0$ vol% and $CH_4 \approx 11$ vol%). The S_N value was similar to the base case.

For case 3 the glycerol feed concentration was reduced to approximately 4 wt%. The H_2 and CO_2 concentration increased compared to the base case, whereas the CO and hydrocarbon concentration were lower, resulting in a slightly more attractive S_N value of 0.8. In the last case (4), a fresh reforming catalyst was used leading to the lowest hydrocarbon concentration and the highest H_2 concentration of all experiments. For instance, C_2H_6, which accounted for ⅓ of the hydrocarbon content in the base case, was not detected in the product gas. The S_N increased to 1.2 which is close to the theoretical maximum of 1.33 (Eq. 3).

5.2. Performance of the integrated process

The results for the integrated process, including methanol synthesis, are presented in Table 4. The equilibria in methanol synthesis were calculated with the data from Table 3 as input and the equilibrium model described in Section 4.3. If applicable, condensation of methanol and water was accounted for in the equilibrium calculations [45]. The equilibrium data should be considered with some care, because the results are based on the assumption of constant gas composition and gas flow from the reformer section. As for the reforming experiments, all methanol synthesis experiments have good closures of the carbon balance (93 – 96%), particularly when regarding the complexity of the integrated process. A detailed summary of the experimental results of the integrated process is given below. As in section 5.1, case 1 is considered as base case and the results of the other experiments are compared to this experiment.

Case	T	H_2	CO	CO_2	C_xH_y	η [a]	MeOH[b]	H_2O	ζ_{CO+CO2}	ζ_c	C_{bal}
	(K)		(vol%)				(wt%)		(%)		
1	468	3	2	44	51	0.27	99	1	58	26	95
equi	468	5	2	44	49	0.28	99	1	59	28	-
2	498	42	0	40	18	0.29	67	33	35	27	96
equi	498	25	1	46	28	0.35	66	34	45	33	-
3	483	49	0	40	11	0.27	65	35	39	26	93
equi	483	15	1	61	23	0.50	65	35	54	48	-
4	481	20	0	60	20	0.62	65	35	71	60	94
equi	481	11	1	69	19	0.65	65	35	70	62	-
Loc.[c]	E	G	G	G	G	F	-	-	-	-	-

[a]Units = (kg methanol/kg glycerol).

[b]The liquid phase is assumed to consist of water and methanol. The methanol concentration here is calculated by 100 wt% – (water concentration). The exact composition of the organic phase is given in Table 5.

[c]Locations where the parameters were measured (see Figure 3).

Table 4. Results of methanol synthesis from glycerol derived syngas [8].

5.2.1. Base case

In the experiment in case 1, the methanol synthesis reactors were operated at 468 K. Hydrocarbons are inert in methanol synthesis and as a results their concentration increased strongly in the off-gas of the methanol synthesis reactor to over 50 vol%. The H_2 and CO concentration in the outlet of the methanol reactor were 3 and 2 vol%, respectively. The CO_2 concentration increased compared to the reforming gas, as mainly CO was converted to methanol. The gas composition and liquid yield at the exit of the methanol reactor were close to equilibrium, with the liquid yield slightly lower and experimental conversion slightly higher than predicted by equilibrium modelling. The overall carbon conversion was 26% which is equal to a methanol yield of 0.27 kg methanol/kg glycerol. The conversion of 26% is the highest conversion possible with such a syngas composition, because equilibrium was reached.

5.2.2. Case 2 – 3

In case 2, a different approach was followed. Due to the Ni based catalyst in the reforming section the hydrocarbon concentration decreased and almost all C_2H_6 was reformed. Furthermore, almost all CO was converted into CO_2, which therefore became the main carbon source of methanol. The temperature of the methanol synthesis reactors was increased compared to the base case (from 468 – 498 K), because methanol synthesis from mainly CO_2 proved to be slower than methanol synthesis from CO. The methanol yield was, with 0.29 kg methanol/kg glycerol, similar to the base case, but in case 2 equilibrium was not reached. Higher conversion are thus possible with longer residence times.

For case 3 the glycerol feed flow was reduced with a factor 3 to improve the gas composition, resulting in a reduction in feed gas flow. Therefore, to obtain higher conversion, the temperature of the methanol synthesis reactor was reduced to 483 K. Again, the carbon conversion and methanol yield (0.27 kg methanol/kg glycerol) were comparable to the base case, but remained far from equilibrium.

5.2.3. Case 4: Long duration experiment in the integrated unit

In case 4 (long duration run, 20 h) an extra methanol synthesis packed bed (P1) was filled with catalyst. The methanol synthesis reactors were operated at three different temperatures (± 518 K (P1), ± 503 K (P2), and 481 – 482 K (P3). The lowest hydrocarbon concentration in the gas phase after the reformer was observed due to the use of a fresh reforming catalyst. As a consequence the corresponding carbon conversion in the methanol synthesis unit increased to 60% (η = 0.62 kg methanol/kg glycerol). Nevertheless, even higher methanol yields are possible as equilibria were not yet achieved. In the first 4 h of the long duration experiment, only the reformer section was operated. Methanol synthesis was carried out over a 16 h period and the hourly liquid methanol yields and volumetric flows at the exit of the methanol synthesis unit (point G in Figure 3) are shown in Figure 16.

Though some scattering in the methanol yield can be noted (due to some pressure fluctuation and uncontrolled release by the back pressure valve during the experiment), the integrated system was running steadily and the methanol yield was more or less constant. The selectivity (σ) is depicted in Figure 16B. The methanol selectivity is equal to the carbon conversion and amounts to 60% on average. An average value of 8.7% of the carbon present in glycerol ends up as CH_4. The scattering pattern for the methanol selectivity is similar to the scattering in the methanol yield in Figure 16A. The selectivity towards CH_4 is also more or less constant. It seems that no deactivation or reduced activity for both the reformer section and the methanol synthesis section were observed during the course of the experiment.

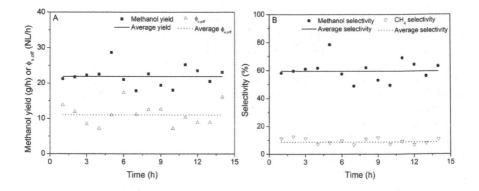

Figure 16. Methanol yields from glycerol and volumetric flow at the exit of the methanol reactor of the long duration experiment (A). Carbon selectivity towards methanol and CH_4 (B) [8].

5.3. Liquid composition after methanol synthesis reactor

The main constituents of the liquid products of the integrated experiments are given in Table 5. The liquid phase was analyzed on methanol, water, and the eight most common HA. In general, the concentration of HA was very low (< 0.23 wt% of the total and < 0.24 wt% of the organic fraction), probably due to the low temperatures of the methanol synthesis and the high CO_2 content of the feed gas, leading to the formation of water [46]. Ethanol was the most predominant among the HA, with a maximum concentration of 1.2 wt‰. Noticeably, when methanol was predominantly synthesized from CO_2 (case 2 – 4) the concentrations of HA were negligible. This is in agreement with literature data which show that the concentrations of HA decrease at higher H_2/CO ratios in the syngas feed [46, 47].

Case	Liquid product				Higher alcohols				
	MeOH	H_2O	HA	Purity[a]	EtOH	2-pro-panol	1-bu-tanol	2-m-1-propanol	2-m-1-butanol
	(wt%)					(wt‰)			
1	97.8	1.1	0.23	99.8	1.2	0.7	0.2	0.2	0.1
2	66.5	33.0	0.00	99.9	0.0	0.0	0.0	0.0	0.0
3	65.9	34.6	0.00	99.9	0.0	0.0	0.0	0.0	0.0
4	65.1	34.9	0.01	99.9	0.1	0.0	0.0	0.0	0.0

[a]Methanol content of the organic fraction.

Table 5. Composition of the liquid phase [8].

5.4. Process analysis

The experiments conducted in the integrated unit were aimed at obtaining high carbon conversions and methanol yields when reforming aqueous glycerol solutions to syngas followed by methanol synthesis. The gas composition after reforming appeared to be the most critical factor and has a major effect on the final methanol yield. Particularly the formation of hydrocarbons should be avoided in the reformer section as hydrocarbons are inert in the subsequent methanol synthesis. Therefore, hydrocarbon reduction was the main objective in the experimental reformer program and was pursued by the application of catalysts, higher reforming temperature, and reduction of the feed concentration. Application of a suitable catalyst (Ni based) indeed led to a considerable reduction in the amount of hydrocarbons in the reformer off-gas, though as a consequence almost all CO was converted into CO_2. Further research will be required to identify reformer catalysts that promote glycerol decomposition rates and hydrocarbon reforming, but do not enhance the WGS reaction. In this respect, Ir-based catalysts are promising because of good performance in aqueous phase reforming [49].

The conversion of CO into CO_2 in the reformer section, as observed when using the Ni based catalyst, is not detrimental for the subsequent methanol synthesis. With the commercial methanol synthesis catalyst used in this study, CO_2 hydrogenation is possible, as was also proven here, though the overall reaction rates in methanol synthesis are lower than in case of CO hydrogenation [50]. An advantage, however, of CO_2 hydrogenation is the high purity of the organic fraction, as the formation of HA is suppressed by, most probably, the presence of water [46, 51].

In the demonstration of the integrated concept, pure glycerol was used as feedstock for the process. When crude glycerin is used salts are present and they have to be removed upfront. Continuous salt removal is possible and has been demonstrated in the literature [52].

6. Conclusion

A successful experimental demonstration of glycerol conversion to methanol was shown by the integration of two processes. Glycerol was reformed in supercritical water to syngas and the syngas was subsequently converted to methanol. Before integration of the two processes the processes were investigated individually. In glycerol reforming a gas containing mainly H_2, CO, CO_2, CH_4, and C_2H_6 was produced. When a Ni based catalyst was used the higher hydrocarbons were reformed and CH_4 approached its equilibrium concentration. In methanol synthesis *in situ* condensation was observed which positively influences the equilibrium conversion. At temperatures around 473 K and pressures above 20 MPa almost complete conversion of the limiting components was obtained.

The methanol yields of the integrated process depended mainly on the gas composition obtained in the glycerol reforming process, which appeared to be the most attractive for methanol synthesis at high temperature and low feed concentration in combination with a Ni based catalyst. The continuous unit was modified during the experimental program to increase the methanol yields. The effluent water of the reformer section was recycled at high pressure, to reduce the water consumption of the process. The highest methanol yield of 0.62 kg methanol/kg glycerol was obtained using the Ni based catalyst in the reformer section and recycling of the effluent water. In this particular experiment glycerol was converted to mainly H_2 and CO_2 and smaller amounts of CH_4 and CO. In this experiment, 60% of the carbon present in the glycerol ends up in methanol. These yields are close to the equilibrium yields. The integrated unit was operated smoothly for more than 16 h without catalyst deactivation.

The scope of the project is much broader than the production of methanol from glycerol for the reuse in biodiesel production. Due to the investigation of the individual processes more insights in reforming and methanol synthesis were obtained. Furthermore, the feedstock for the reforming process was glycerol in this case, but several types of biomass (preferably liquid) including aqueous phase fractions from pyrolysis oil upgrading, black liquor, etc. can be used for the reforming process. When these types of feedstocks are 'green' renewable methanol can be produced, which is a promising process for the (near) future.

Acknowledgements

The authors acknowledge the assistance given by the following companies and publishers in permitting the reproduction of figures, tables, and portions of the text (with minor adaptations) from their publications.

Figure 1 was published with courtesy of Desmet Ballestra [16]. The data has been obtained from the European Biodiesel Board.

Figures 3, 10, and 16 and Tables 2 – 5, and large portions of Sections 1, 2, and 5 were originally published in Chemical Engineering Journal [8] and used with permission.

Figure 4 [19] is reproduced in modified form by permission of The Royal Society of Chemistry (http://dx.doi.org/10.1039/B810100K).

Figures 5 and 8 and portions of Section 3 were originally published in the Journal of Supercritical Fluids [29] and used with permission.

Figure 6 and portions of Section 3 were originally published in the Journal of Supercritical Fluids [18] and used with permission.

Figures 12 and 13B and portions of Section 4 were originally published in Chemical Engineering Science [41] and used with permission.

Figures 13A and 14 were reprinted with permission from [45], copyright 2012 American Chemical Society.

The authors would like to acknowledge the EU for funding the work through the 5th and the 7th Framework Program (Superhydrogen: ENK5-2001-00555 and Supermethanol: no. 212180) and Agentschap NL (EOSLT05020). The initial screening for Supermethanol was funded by Agentschap NL (no. 0268-05-04-02-011 and no. NEOT01008). We also would like to thank the partners of the Supermethanol project, and in particular D. Assink, K.P.J. Lemmens, S.D.G.B. Nieland from BTG, The Netherlands for building and operating the units, J. Vos from BTG for proofreading, and E. Wilbers, J.G.M. Winkelman and J. Bos from the University of Groningen, The Netherlands for their assistance in the work concerning the methanol synthesis and J. Bos for taking the pictures of the reactor. V.A. Kirillov from the Boreskov Institute of Catalysis, Russia is acknowledged for supplying the reforming catalyst.

Author details

Joost G. van Bennekom[1], Robertus H. Venderbosch[2] and Hero J. Heeres[1*]

*Address all correspondence to: h.j.heeres@rug.nl

1 University of Groningen, Green Chemical Reaction Engineering, Groningen, The Netherlands

2 BTG, Biomass Technology Group, Enschede, The Netherlands

References

[1] Olah GA, Goeppert A, Prakash GKS. Beyond oil and gas: The methanol economy. Weinheim: Wiley-VCH; 2006.

[2] Beenackers AACM, Swaaij WPM van. Methanol from wood II. Current research and development programs. Int. J. Sol. Energy. 1984; 2(6) 487-519.

[3]   Beenackers AACM, Swaaij WPM van. Methanol from wood I. Process principles and technologies for producing methanol from biomass. Int. J. Sol. Energy. 1984; 2(5) 349-367.

[4]   Xiuli Y. Synthesizing methanol from biomass-derived syngas. Hong Kong: The university of Hong Kong; 2005.

[5]   Norbeck JM, Johnson K. Evaluation of a process to convert biomass to methanol fuel. In: Agency, U. S. E. P., editor. Cincinnati2000.

[6]   Ekbom T, Lindblom M, Berglin N, Ahlvik P. Cost-competitive, efficient biomethanol production from biomass via black liquor gasification (Nykomb Synergetics AB, 2003).

[7]   Knoef HAM. editor Handbook Biomass Gasification. Enschede: BTG Biomass Technology Group; 2005.

[8]   Bennekom JG van, Venderbosch RHV, Assink D, Lemmens KPJ, Heeres HJ. Bench scale demonstration of the Supermethanol concept: The synthesis of methanol from glycerol derived syngas. Chem. Eng. J. 2012; 207-208 245-253

[9]   Borgwardt RH. Methanol production from biomass and natural gas as transporation fuel. Ind. Eng. Chem. Res. 1998; 37 3760-3767.

[10] Chmielniak T, Sciazko M. Co-gasification of biomass and coal for methanol synthesis. Appl. Energy. 2002; 74 393-403.

[11] Hamelinck CN, Faaij APC. Future prospects for production of methanol and hydrogen from biomass. J. Power Sources. 2002; 111 1-22.

[12] Li H, Hong H, Jin H, Cai R. Analysis of a feasible polygeneration system for power and methanol production taking natural gas and biomass as materials. Appl. Energy. 2010; 87 2846-2853.

[13] Rens GLMA van, Huisman GH, Lathouder H de, Cornelissen RL. Performance and exergy analysis of biomass-to-fuel plants producing methanol, dimethylether or hydrogen. Biomass Bioenergy. 2011; 35 S145-S154.

[14] Xu Y, Ye TQ, Qiu SB, Ning S, Gong FY, Liu Y, Li QX. High efficient conversion of CO_2-rich bio-syngas to CO-rich bio-syngas using biomass char: A useful approach for production of bio-methanol from bio-oil. Bioresour. Technol. 2011; 102 6239-3245.

[15] Directive 2009/28/EC of the European Parliament and of the Council of 23 April 2009 on the promotion of the use of energy from renewable sources and amending and subsequently repealing Directives 2001/77/EC and 2003/30/EC. Official Journal of the European Union. Strasbourg 2009. p. 47.

[16] Greyt W de. Introduction on glycerol as co-product from biodiesel plants. Innovative uses of glycerol from the biodiesel process; 2011; Brussels.

[17] Bennekom JG van, Vos J, Venderbosch RH, Torres MAP, Kirilov VA, Heeres HJ, Knez Z, Bork M, Penninger JML. Supermethanol: Reforming of crude glycerine in su-

percritical water to produce methanol for re-use in biodiesel plants. 17th European Biomass Conference and Exhibition; 2009; Hamburg.

[18] Bennekom JG van, Venderbosch RH, Assink D, Heeres HJ. Reforming of methanol and glycerol in supercritical water. J. Supercrit. Fluids. 2011; 58(1) 99-113.

[19] Peterson AA, Vogel F, Lachance RP, Fröling M, Jr. MJA, Tester JW. Thermochemical biofuel production in hydrothermal media: A review of sub- and supercritical water technologies. Energy Environ. Sci. 2008; 1 32-65. http://dx.doi.org/10.1039/B810100K

[20] Toor SS, Rosendahl L, Rudolf A. Hydrothermal liquefaction of biomass: A review of subcritical water technologies. Energy. 2011; 36 2328-2342.

[21] Gadhe JB, Gupta RB. Hydrogen production by methanol reforming in supercritical water: suppression of methane formation. Industrial and Engineering Chemistry Research. 2005; 44(13) 4577-4584.

[22] Brunner G. Near critical and supercritical water. Part I. Hydrolytic and hydrothermal processes. J. Supercrit. Fluids. 2009; 47 373-381.

[23] Elliot DC. Catalytic hydrothermal gasification of biomass. Biofuels, Bioprod. Biorefin. 2008; 2 254-265.

[24] Kruse A. Supercritical water gasification. Biofuels, Bioprod. Biorefin. 2008; 2 415-437.

[25] Kruse A. Hydrothermal biomass gasification. J. Supercrit. Fluids. 2009; 47 391-399.

[26] Matsumura Y, Minowa T, Potic B, Kersten SRA, Prins W, Swaaij WPMv, Beld Bvd, Elliot DC, Neuenschwander GG, Kruse A, Jr. MJA. Biomass gasification in near- and supercritical water: Status and prospects. Biomass Bioenergy. 2005; 29 269-292.

[27] Basu P, Mettanant V. Biomass gasification in supercritical water - A review. Int. J. Chem. Reactor Eng. 2009; 7 1-61.

[28] Weingärtner H, Franck EU. Supercritical water as a solvent. Angew. Chem. Int. Ed. 2005; 44 2672-2692.

[29] Bennekom JG van, Kirillov VA, Amosov YI, Krieger T, Venderbosch RH, Assink D, Lemmens KPJ, Heeres HJ. Explorative catalyst screening studies on reforming of glycerol in supercritical water. J. Supercrit. Fluids. 2012; 70 171-181.

[30] Mittasch A, Winkler K, Pier M. Verfahren zur Gewinnung organischer Verbindungen durch katalytische Gasreaktionen. DE441433, Germany, 1923.

[31] Skrzypek J, Stoczynski J, Ledakowicz S. Methanol synthesis. Warsaw: Polish Scientific Publishers PWN; 1994.

[32] Supp E. How to produce methanol from coal. Berlin: Springer-Verlag; 1990.

[33] Reubroycharoen P, Vitidsant T, Asamic K, Yoneyama Y, Tsubaki N. Accelerated methanol synthesis in catalytically active supercritical fluid. Catal. Commun. 2003; 4(9) 461-464.

[34] Westerterp KR, Kuczynski M, Kamphuis CHM. Synthesis of methanol in a reactor system with interstage product removal. Ind. Eng. Chem. Res. 1989; 28 763-770.

[35] Kuczynski M, Oyevaar MH, Pieters RT, Westerterp KR. Methanol synthesis in a countercurrent gas-solid-solid trickle flow reactor, an experimental study. Chem. Eng. Sci. 1987; 42(8) 1887-1898.

[36] Haut B, Halloin V, Amor HB. Development and analysis of a multifunctional reactor for equilibrium reactions: Benzene hydrogenation and methanol synthesis Chem. Eng. Process. 2004; 43 979-986.

[37] Castier M, Rasmussen P, Fredenslund A. Calculation of simultaneous chemical and phase equilibria in nonideal systems. Chem. Eng. Sci. 1989; 44(2) 237-248.

[38] Hansen JB, Joensen F, editors. High conversion of synthesis gas into oxygenates. Proceeding of the NGCS; 1991; Oslo: Elsevier Science Publishers B.V.; 1990.

[39] Sorensen EL, Perregaard J. Condensing methanol synthesis and ATR - The technology choice for large-scale methanol production. Stud. Surf. Sci. Catal. 2004; 147 7-12.

[40] Topsoe HFA, Hansen JB. Method of preparing methanol. US5262443, United States 1993.

[41] Bennekom JG van, Venderbosch RH, Assink D, Lemmens KPJ, Winkelman JGM, Wilbers E, Heeres HJ. Methanol synthesis beyond chemical equilibrium. Chem. Eng. Sci. 2013; 87 204-208

[42] Poling BE, Prausnitz JM, O'Connel JP. The properties of gases and liquids. 5th ed. Singapore: McGraw-Hill; 2007.

[43] Mathias PM. A verstatile equilibrium equation of state. Ind. eng. Chem. Process Des. Dev. 1983; 22 385-391.

[44] Graaf GH, Sijtsema PJJM, Stamhuis EJ, Joosten GEH. Chemical equilibria in methanol synthesis. Chem. Eng. Sci. 1986; 41 2883-2890.

[45] Bennekom JG van, Winkelman JGM, Venderbosch RH, Nieland SDGB, Heeres HJ. Modeling and experimental studies on phase and chemical equilibria in high pressure methanol synthesis. Ind. Eng. Chem. Res. 2012; 51(38) 12233-12243

[46] Denise B, Sneeden RPA. Hydrocondensation of carbon dioxide: IV. J. Mol. Catal. 1982; 17 359-366.

[47] Smith KJ, Anderson RB. A chain growth scheme for the higher alcohols synthesis. J. Catal. 1984; 85(2) 428-436.

[48] Subramani V, Gangwal SK. A review of recent literature to search for an efficient catalytic process for the conversion of syngas to ethanol. Energy Fuels. 2008; 22 814-839.

[49] Davda RR. A review of catalytic issues and process conditions for renewable hydrogen and alkanes by aqueous-phase reforming of oxygenated hydrocarbons over supported metal catalysts. Appl. Catal., B. 2005; 56 171-186.

[50] Klier K, Chatikavanij V, Herman RG, Simmons GW. Catalytic synthesis of methanol from CO/H$_2$. IV. The effects of carbon dioxide J. Catal. 1982; 74 343-360.

[51] Forzatti P, Tronconi E, Pasquin I. Higher alcohol synthesis. Catal. Rev. Sci. Eng. 1991; 33(1&2) 109-168.

[52] Schubert M. Catalytic hydrothermal gasification of biomass - Salt recovery and continuous gasification of glycerol solutions. Zürich: ETH; 2010.

Chapter 2

Utilization of Crude Glycerin from Biodiesel Production: A Field Test of a Crude Glycerin Recycling Process

Hayato Tokumoto, Hiroshi Bandow,
Kensuke Kurahashi and Takahiko Wakamatsu

Additional information is available at the end of the chapter

1. Introduction

1.1. Background

Worldwide, increasing quantities of biodiesel fuel (BDF) are produced, along with bioethanol. The production of BDF generates glycerin (also known as glycerol) as a by-product. Because alkaline metal oxides or alkaline hydroxides are commonly used as catalysts for the transesterification of vegetable oils with methanol, the glycerin stream is strongly alkaline, and therefore must be neutralized, demineralized, rinsed with water, and dried before glycerin is combusted. This has been a barrier to the popularization of BDF [1-3].

Osaka Prefecture University (abbreviated as OPU) in Sakai, Japan, is promoting a "Campus Zero Emissions" project, intended to recycle the resources within the campus. We have a bench-scale methane fermentation plant and BDF production plant based on ultrasound that is capable of producing BDF and methane from waste cooking oil [4-7] and food waste.

Methane fermentation is one of the main processes used for food waste (nitrogen and carbon mixtures) stabilization. High nitrogen concentration and pH inhibits growth of bacteria because of toxicity caused by high levels of ammonia. Microorganisms require carbon and nitrogen for metabolism, and the relationship between their amounts in organic materials is represented by the C/N ratio. Optimum C/N ratios in anaerobic digesters are between 20 and 30. A high C/N ratio is an indication of rapid consumption of nitrogen by methanogens, and results in lower gas production. On the other hand, a lower C/N ratio causes ammonia accumulation and pH values exceeding 8.5, which is toxic to methanogenic bacteria. Optimum C/N ratios in digester contents can be achieved by mixing materials with high and low C/N ratios, such as the raw glycerin byproduct of BDF production.

One of the authors, Tokumoto, has filed three patent applications regarding a leading-edge technology that allows the fermentation of strongly alkaline waste glycerin using anaerobic microorganisms, without additional processes. This technology is expected to compete with fermentation technology that uses a microorganism with high glycerin degradation ability. However, the latter is, in general, based on the selection of the most favorable individually cultivated microorganism system from a wide variety of individually isolated fungi. The cost of an operation that utilizes such delicate microorganisms is high and this is one of the barriers to commercialization of the system (Figure 1) [8-12]. Using high-level experience in microbiological control, our project is working towards the establishment of a new process based on a low-cost, combined cultivation system.

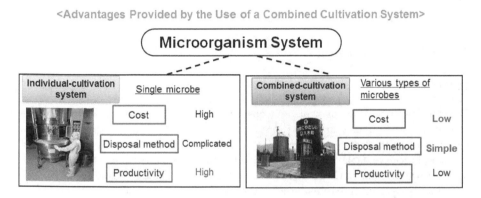

Figure 1. Comparison between individual- and combined-cultivation systems

1.2. Purpose

The campus zero-emission project is intended to establish the business model for OPU's resource recycling process using waste cooking oil. In this model, we produce BDF from waste cooking oil discharged from the dining halls on campus and generate methane from the glycerin by-product to supply fuel for motorcycles, vehicles, and electric power facilities [13-14]. This research activity will result in verification tests based on about 10,000 students and teaching/clerical staff members as monitors, and then develop the test results into a comprehensive recycling process for waste cooking oil.

Commercial-scale plants will be standardized at the end of this project, based on the operational data from the methane fermentation plant and utilization facilities (motorcycles, other vehicles, and electric power facilities) shown in the figure below.

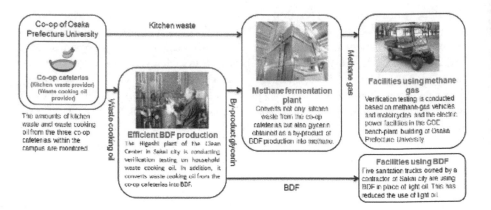

Figure 2. Entire experiment flow chart

2. Results and discussion

2.1. Methane fermentation reactor

The methane fermentation plant has only one fermenter, with a fermentation vessel volume of 10 m^3 and an operating capacity of 5 m^3. From startup, the plant used bean curd lees as a fermentation substrate (fermentation feedstock for breeding microorganisms and generating methane) to achieve stable operation. We then added another fermentation substrate – part of the kitchen waste from the OPU's co-op cafeteria. After stable operation was achieved, we then added waste glycerin, the by-product generated during BDF production.

2.2. Operational testing on the fermentation plant

2.2.1. Alkali considerations of the microorganism fermentation plant

The figure below shows the correlation between the amount of gas generated and the raw material disposal rate.

We purchased seed sludge from the methane fermentation plant of the Bioecology Center in Yagi Cho, Kyoto Prefecture and charged this into the fermenter in our methane fermentation plant. After the startup (neutralization) process, we used bean curd lees as a fermentation substrate, which resulted in favorable biogas production after about 10 days (Figure 3). In this case, we observed a positive correlation between the amounts of bean curd lees disposed of and the biogas generated. From May 12 to June 5, the amount of generated biogas per unit amount of bean curd lees was 0.68 m^3/kg, calculated by the least-squares method (Figure 3).

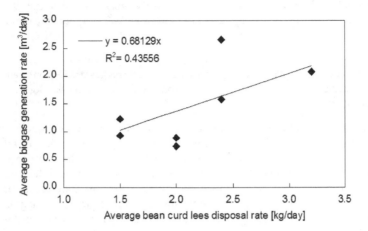

Figure 3. Correlation between the amounts of gas generated and bean curd lees disposal rate

2.2.2. Biogas from co-op waste

Temporal changes in the amounts of biogas produced after the kitchen waste from the co-op cafeteria was added are shown in Figure 4.

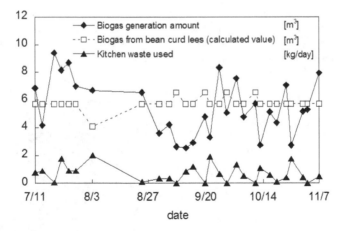

Figure 4. Amount of biogas generated from kitchen waste discharged from the co-op cafeteria

The amounts of biogas generated shown in Fig. 4 are actual measured values. The amounts of biogas from the bean curd lees were obtained by multiplying the biogas generation rates calculated in Figure 3 by the amount of bean curd lees. The subtraction of these values may result in negative values in some cases. For this reason, the figure does not indicate the amounts of biogas from the kitchen waste discharged from the co-op cafeteria, but only the amounts of waste added.

In producing these results, we successfully disposed of co-op waste without destabilizing the biogas production at the plant, which provided a favorable biogas production even using co-op kitchen waste. In this case, we successfully converted a maximum of 2 kg co-op kitchen waste into biogas per day.

The results, however, also indicated that the kitchen waste delivered a conversion efficiency lower than that for bean curd lees alone. Bean curd lees are vegetable-protein food and contain significant nitrogen and exhibit a low C/N value. The C/N value is usually regarded as an index of the favorability of substrates towards anaerobic fermentation [15], and the low value for bean curd lees suggests that the existence of the low C/N-value substrates, bean curd lees, in this system could decrease the reactivity of the microorganisms producing biogas. It should be noted, however, that in the long-term, increases in the use of co-op waste lead to gradual increases in the amount of biogas generated.

Figure 5 shows the monthly number of cafeteria users in 2011. A university-specific trend was observed, in that the number of users decreased during the long vacation period in August and September. A similar trend among the kitchen waste used and amount of biogas generated in Figure 4 and the number of the cafeteria users in Figure 5 indicates that the gas production amount is strongly correlated with the number of cafeteria users. Consequently, the largest influence is the decrease in the amount of organic substances contained in the kitchen waste. When the number of users was stable in September and October, the biogas production stabilized, meaning that the plant seems to provide stable operation as a whole. With these behaviors in mind, this research is characterized by the fact that investigating the amount of waste enables the estimation of the performance of the fermentation plant, based on the number of users of the dining facility or facilities in the business place.

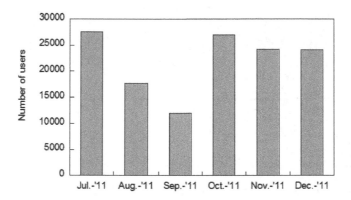

Figure 5. Number of co-op cafeteria users

2.2.3. *Converting waste glycerin into biogas*

The amounts of biogas achieved when bean curd lees, kitchen waste, and waste glycerin were used as substrates are shown in Figure 6.

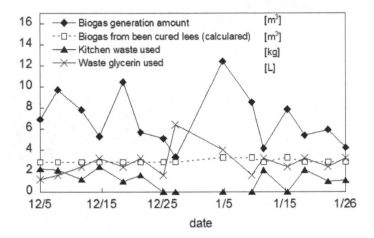

Figure 6. Amounts of biogas generated from bean curd lees, kitchen waste, and waste glycerin

The amounts of biogas shown are actual measured values. The amounts of biogas produced from the bean curd lees were obtained by multiplying the biogas generation rates calculated in Figure 3 by the amount of bean curd lees. For the kitchen waste and waste glycerin, the figure indicates the amounts used.

The amount of waste glycerin is correlated with the amount of biogas generated, whereby the addition of glycerin significantly increases the amount of biogas. Therefore the disposal of waste glycerin in this process has significant advantages. It is estimated that the biogas generation rate per unit amount of glycerin was 1.63 m^3/kg on average. This is about 2.5 times the biogas generation rate when bean curd lees alone were used. It was also observed that about 2 kg of co-op kitchen waste per day were steadily decomposed by fermentation, as seen above. The operation continued steadily for about a month. In addition, the amounts of bean curd lees, kitchen waste, and waste glycerin we used this time were small compared with the operating capacity of the fermentation vessel, and the full capacity has not yet been used.

2.2.4. *Estimated reduction in CO_2 emissions*

The disposal flow and the calculated reduction in CO_2 emissions per unit amount disposed are shown in Figure 7.

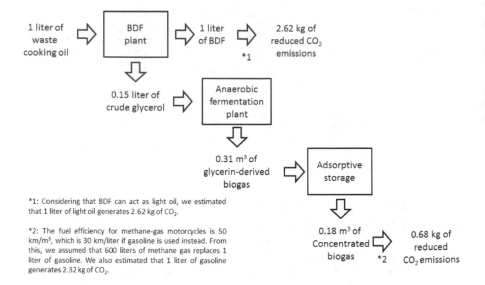

*1: Considering that BDF can act as light oil, we estimated that 1 liter of light oil generates 2.62 kg of CO_2.

*2: The fuel efficiency for methane-gas motorcycles is 50 km/m³, which is 30 km/liter if gasoline is used instead. From this, we assumed that 600 liters of methane gas replaces 1 liter of gasoline. We also estimated that 1 liter of gasoline generates 2.32 kg of CO_2.

Figure 7. Disposal flow and reduced CO_2 emissions

Because waste cooking oil is a plant-derived organic substance, it is possible to estimate the reduction in CO_2 emissions based on carbon neutrality. For simplicity, we ignored the CO_2 emissions associated with machine operation. We also assumed that we could treat BDF equally as an alternative to light oil. Furthermore, we considered that 600 liters of biogas can replace one liter of gasoline, based on the fuel efficiency of motorcycles. The definitions of CO_2 emissions from fossil fuels were based on the guidelines from the Ministry of Environment [16].

The disposal of glycerin at the anaerobic fermentation plant effectively reduces CO_2 emissions by more than 30% compared with waste cooking oil wholly incinerated, without additional treatment. When glycerin is incinerated, a fossil fuel is normally used as a combustion aid. If this is taken into consideration, this method may have an even greater effect on reducing CO_2 emissions. Only part of the kitchen waste from the co-op cafeteria is currently disposed of. Disposal of all kitchen waste will further reduce CO_2 emissions.

While a long-term testing and verification period is required, it is expected that biogas generation from waste glycerin will further improve, providing a larger reduction in CO_2 emissions.

2.3. Biogas fuel conversion facilities

The effective use of the biogas generated in the methane fermentation vessel as a fuel requires a system for temporarily storing the biogas, a system for concentrating the methane contained in the biogas, and facilities that can make effective use of the concentrated (refined) biogas.

The concentrated methane gas can be used "as-is", if the required pressure is low. On the other hand, if it is used as a fuel for vehicles, then equipment for charging biogas into the vehicles is required. Figure 8 shows the biogas use flow chart. A description of the equipment we operated in this research is given below.

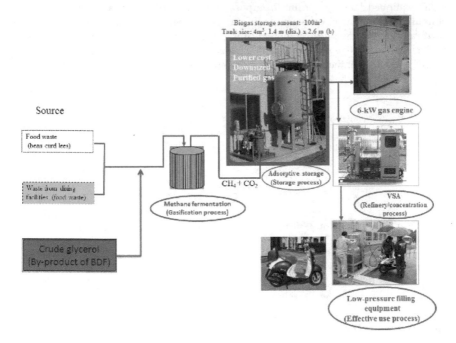

Figure 8. Biogas use flow chart

2.3.1. Adsorptive biogas storage system

2.3.1.1. Principle

The effective use of the biogas generated by methane fermentation requires a buffer tank for storing gas in order to make 100% use of the biogas, because it is difficult to balance the amount generated with usage at any one time.

In most cases currently, biogas is stored under a low pressure of a few hundred mm of H_2O, which is the gas pressure inside the fermentation vessel. However, this method can store only the same amount of gas as a gas tank. The storage of a larger amount of gas involves problems associated with size and equipment cost. To solve these problems, a method has been considered that fills the gas tank with a microporous absorbent to enable adsorptive storage of gas, allowing the storage of a large amount of gas at room temperature and under a relatively low pressure [17].

This method uses a phenomenon whereby methane, the major component of biogas, is physically absorbed in micropores of absorbents at a density close to that of its liquid state (Figure 9). This technology is expected to provide large-volume storage even under a relatively low pressure, because it even absorbs methane that is not liquefied by pressure. This new, attractive storage method, if commercialized, could provide safer storage of digestion gas with a lightly equipped device, according to purpose, and allow the transportation of biogas to other points of consumption, which is not common at the moment. In addition, it has the advantage that it can provide gas purification, which is required for effective use of biogas, through simultaneous adsorptive storage. Because biogas must be effectively stored in a limited area within the premises of the plant at this time, we used an adsorptive methane storage system.

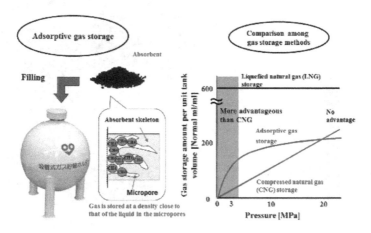

Figure 9. Principle of adsorptive storage

2.3.1.2. Device specifications (stationary large-capacity type, Figure 10)

- Type

  Biogas-absorbent-filling-type, vertical cylindrical pressure tank
- Applicable laws and regulations

  Construction Code for Pressure Vessels (second-class pressure vessels)
- Intended use

  Adsorptive storage of biogas
- Gas storage capacity

  Effective amount: approx. 100 Nm^3

  activated carbon storage capacity of about 4 m^3
- Dimensions

1,700 mm (dia.) x 2,600 mm (H) (Trunk: 1,900 mm)

- Fluid

  Refined biogas (CH_4: 60%, CO_2: 40%)

- Operating pressure

  0 – 0.75 MPa (G)

- Operating temperature

  10 – 50°C

- Design temperature

  50°C

- Design pressure

  0.8 MPa (G)

- Absorbent amount

  Approx. 2.1 tons (coconut shell activated carbon)

- Main unit

  SS400 (epoxy-coated inner surface)

- Ladder and handrail

  SS400 and SGP

- Hanging ring

  SS400

- Foundation bolts and nuts

  SS400

Figure 10. The adsorptive biogas storage system

2.3.1.3. Storage performance

The storage performance depends significantly on the ambient air temperature, absorbent temperature, and the incoming and outgoing flow rates of the target gas, and is therefore difficult to measure accurately from the actual tank dimensions (Figure 10). For this reason, we used a small insulating container to measure the storage performance, similar to that of the actual tank, and evaluated the storage performance by rapidly filling it with 100% methane. We charged the gas at an absorbent temperature of 25°C. It was verified that, with the activated carbon charged 100%, the container can store 100 m³ of methane, approximately 25 times the tank capacity, if it operates in the range of 0 – 0.6 MPa.

2.3.2. Adsorptive, isolated methane concentrating device

2.3.2.1. Principle

In general, biogas generated in the methane fermentation vessel contains CH_4, CO_2, saturated H_2O, H_2S, and trace quantities of organic components generated during the decomposition process. To make effective use of this gas, it is necessary to refine it (concentrate the methane, which will act as the fuel). If it is used as fuel for vehicles, the methane must be delivered at a purity of at least 95% and the water vapor must have a dew point of -55°C or less.

The methods for refining biogas include pressure swing adsorption using an absorbent, the separation membrane technique using polymeric separation membranes, and the absorption technique using (alkaline) water. In this research, we used pressure swing adsorption (PSA) because it is able to produce methane and remove water [18].

As Figure 11 shows, the device uses an absorbent with a controlled micropore diameter to selectively absorb and remove carbon dioxide based on the difference in their molecular sizes. It also concentrates the methane and removes impurities. This means that using an absorbent with micropore sizes between the molecular sizes of methane and carbon dioxide, the device can separate methane, the major component of biogas, as well as carbon dioxide, water and impurities. At the same time, it also absorbs and removes water, which has a smaller molecular diameter than carbon dioxide. Note that the molecular sizes follow the order methane > carbon dioxide > water.

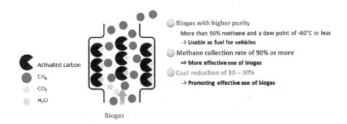

Figure 11. Principle of biogas purification

Using two adsorption towers filled with the absorbent described above, concentrated methane with a low dew point can be seamlessly obtained by alternately repeating the adsorption and regeneration processes.

2.3.2.2. Device specifications (Figure 12)

i.   Entrance biogas conditions

      1.   Flow rate: 3.2 Nm^3/h

      2.   Pressure: 100 mm H_2O

      3.   Biogas composition

      CH_4: 60%, CO_2: 40%, water: saturated

      Trace quantities of ingredients (e.g., hydrogen sulfide and ammonia): 1 ppm or less

ii.   Exit product gas conditions

      1.   Methane purity: at least 95%

      2.   Methane collection rate: at least 90%

Figure 12. The adsorptive, isolated methane concentrating device

2.3.2.3. Characteristics of methane concentration

The absorbent used for methane concentration is a carbon molecular sieve (CMS) with a micropore diameter adjusted to be approximately 0.3 to 0.35 nm [19]. If it is used to separate carbon dioxide and methane, then the difference in adsorption rate is used instead of the difference in equilibrium adsorption capacity. Table 1 and Figure 13 show the data for the equilibrium amount adsorbed and adsorption rate curve, respectively [20]. As Figure 13 indicates, the amount of carbon dioxide reached 90% of the equilibrium adsorption capacity within one minute, while almost no methane was absorbed. This principle can be used to perform adsorption separation.

	CO_2	Methane
Adsorption amount (ml/g)	55.2	26.9

Table 1. Equilibrium adsorption capacities (under one atmosphere pressure)

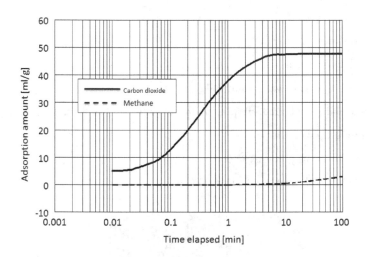

Figure 13. Adsorption rate curves for carbon dioxide and methane

2.3.3. Operational testing for biogas

1. Overview

Refined biogas can be effectively used as a fuel in vehicles and cooking appliances. In the plant, we verified the establishment of effective systems that allow biogas to be used as a fuel in light cars for food sales, service buggy cars used within the campus and for motorcycles, as well as a fuel for cooking in the cafeteria and other facilities.

Using a filler, we charged approximately 95% biogas refined through PSA into a typical natural-gas light car, on-campus service buggy cars, motorcycles equipped with a fuel canister filled with an absorbent, and adsorptive storage cylinders for transfer filled with an absorbent. The filling equipment used charges gas under a low filling pressure of 0.98 MPa or less, and is not, therefore, restricted by any laws or regulations in Japan.

As methane vehicles, we used on-campus light minivans, on-campus adsorptive service buggy cars, and adsorptive motorcycles. Large quantities of biogas must be stored under a high pressure. This requires adherence to the High Pressure Gas Safety Act and other laws and regulations, causing the unit price of biogas to rise. The system verified in this research project eliminates the need to address this matter, providing safe transfer of large volumes of biogas.

Figure 14 shows the flow chart for extracting methane from a methane adsorptive storage tank for use as a fuel in vehicles. Using these systems, biogas is expected to be able to be used in a wider range of applications, including the consumer segment.

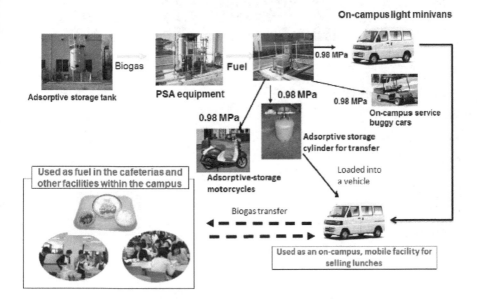

Figure 14. Flow chart for the use of biogas as a fuel in vehicles and cooking

2. Overviews of the devices

[Biogas filler] (Figure 15)

- Type

  Low-pressure filling equipment for biogas

- Discharge flow rate of the gas compressor

  1.6 Nm^3/h or more

- Filling pressure

  0.98 MPa G or less

- Entrance biogas composition

  Methane: 95%, CO_2: 5%

Figure 15. The biogas filler

[Light minivan] (Figure 16)

- Fuel tank
- Storage pressure

  0.98 MPa
- Storage capacity

  Approx. 0.5 m^3
- Travel range

  6 – 8 km

Figure 16. A light minivan

[On-campus adsorptive service buggy (Figure 17)

- Fuel tank
- Storage pressure

  0.98 MPa

- Storage capacity

  3.5 m^3

- Travel range

  70 – 90 km

Figure 17. An on-campus adsorptive service buggy

[Adsorptive Motorcycle] (Figure 18)

- Fuel tank

- Storage pressure

  0.98 MPa

- Storage capacity

  1 m^3

- Travel range

  50 km

Figure 18. An adsorptive motorcycle

3. Conclusion

3.1. Consideration of business sizes

	Motorcycle	Buggy	Vehicle (within the premises)	Vehicle (outside the premises)
Biogas storage capacity [m^3]	1	3.5	0.5	12.5
Mileage [km]	50	80	7	175
Mileage per day [km]	5	10	7	50
Annual number of operation days [days]	200	200	200	200
Required monthly number of cafeteria users [people]	506	2212	2528	18057
Annual disposal amount of glycerin [L]	9.74	42.6	48.7	348
Annual reduction of CO$_2$ emissions [kg] — BDF	170	744	850	6075
Annual reduction of CO$_2$ emissions [kg] — Biogas	44.1	193	221	1577
Annual reduction of CO$_2$ emissions [kg] — Total	214	937	1071	7651
Use	Delivery of mail, etc.	Travel within the premises	Material transportation within the premises	Material transportation from/to the premises

Table 2. Utilizations of biogas and their effects on CO$_2$ emissions reduction

Based on numerical values regarding biogas uses and utilizations (from product catalog data), we considered the required business sizes.

In the model, we used the number of users of the co-op cafeteria and the amount of waste cooking oil generated. The co-op cafeteria of Osaka Prefecture University is used by 23,000 people per month on average and 3,000 liters of waste cooking oil are discharged annually. We considered the amount of glycerin derived from BDF production to be one quarter of the amount of waste cooking oil. Table 2 summarizes the uses and utilizations of the biogas along with their effects on CO$_2$ emissions reduction.

Based on the data in the table, we estimated that the business model proposed by this research can be applied to any business place that has a dining facility used by hundreds of people a month. With increases in the number of users, the form of use and utilization develops; if a business place has a dining facility or facilities used by more than 20,000 people a month, then it is expected that the business can expand the use range to include the utilization of business vehicles.

Brazil, an excellent exemplar for biomass energy power generation, has started to make efforts to reduce fossil fuel use by blending 10% BDF into light oil, similar to the use of bioethanol in the past. The process proposed can be applied to all vehicles, including natural-gas and diesel vehicles, as long as they use an internal combustion engine. The largest challenge is fuel storage. Because gas changes volume with temperature, it is important to increase the amount of biogas stored per unit volume. Our partner, Osaka Gas Engineering, owns leading-edge technology for biogas storage and its application, which is expected to popularize the business model.

3.2. Future challenges

Below are the future challenges associated with business projects that use biogas as a fuel and other uses:

1. Decreasing the fuel price per unit heating value to or below that of city gas

    In many cases, the use of biogas as a fuel is compared with the use of city gas in terms of cost because they exhibit similar properties [21]. The comparison, however, normally indicates that biogas has no clear advantage.

    On the other hand, to achieve sustainable development, it is important to use biogas – a recyclable, carbon-neutral fuel. For this reason, it is necessary to give a preferential tax rate according to its use and implement a system that facilitates subsidies for equipment installation, for example.

2. Developing a comprehensive plan covering the entire surrounding area when constructing a biogas generation facility

    To use biogas, a waste-derived fuel, at low cost, it is imperative that the raw material (waste) can be collected intensively and that local facilities can use the generated biogas. This means that it is necessary to develop a comprehensive plan covering all neighboring areas when constructing a biogas generation facility.

    Based on the characteristics of the university, this research covers all processes ranging from the generation of waste and the production of biogas and BDF to their uses, so it may provide an excellent case study for developing a regional plan.

3. Eliminating restrictions to the use of biogas through fuel transfer, based on an adsorptive storage technology

    Currently, the sewage plants in Osaka city use sludge digestion to dispose of sludge. For the effective use of the biogas generated there, electric power generation and many other applications are being considered and implemented.

    On the other hand, when sewage plants and other facilities make effective use of biogas, that use is subject to many laws and regulations (e.g., High Pressure Gas Safety Act, Gas Business Act, and Building Standards Act), depending on the installation site, and therefore is restricted in some cases. If electric power is generated within the premises of a sewer plant, the generation efficiency is lower than that of large electric power generation facilities, and the location where the collected hot water should be used must be considered.

The biogas transfer system based on the adsorptive storage technique is an effective solution to these problems.

Examples of useful applications may include a sewer plant or a methane fermentation facility that cannot make effective use of biogas because it is located in a non-industrial area; biogas generated there can be transferred to a large electric power generation facility using an adsorptive storage tank installed in an ISO-defined container for use as a fuel. In this case, the use of biogas is not subject to the various laws and regulations and it is possible to safely transfer large volumes of biogas. Biogas has similar properties to natural gas and can be used at power plants and other facilities that use natural gas as fuel. Therefore, its use is expected to grow.

Acknowledgements

This research was conducted with the considerable help of Professor Taketoshi Okuno, President of Osaka Prefecture University, and Professor Masakazu Ampo, Vice-President of the university. In addition, the research was subsidized by Osaka City as a project to help verify the practicality of environmental and energy-related technologies. We would like to express our heart-felt thanks to all of the above.

Author details

Hayato Tokumoto[1*], Hiroshi Bandow[1], Kensuke Kurahashi[2] and Takahiko Wakamatsu[3]

*Address all correspondence to: tokumoto@chemeng.osakafu-u.ac.jp

1 Department of Chemical Engineering, Osaka Prefecture University, Gakuen-cho, Sakai, Osaka, Japan

2 Osaka Prefecture University College of Technology, Saiwai, Neyagawa, Osaka, Japan

3 Energy and Environment Business Div, Energy Business and Engineering Dept, Osaka Gas Engineering Co., Ltd., Japan

References

[1] Yazdani, S.S, Gonzalez, R. Anaerobic fermentation of glycerol : a path to economic viability for the biofuels industry. Current Opinion in Biotechnology 2007;18(3) 213-219.

[2]   Choi W J. Glycerol-Based Biorefinery for Fuels and Chemicals. Recent Patents on Biotechnology 2008;2(3) 173-180.

[3]   Siles López J.A, Martín Santos M.d.l.A, Chica Pérez A.F, Martín Martín A. Anaerobic digestion of glycerol derived from biodiesel manufacturing. Bioresource Technology 2009;100(23) 5609-5615.

[4]   Gunaseelan, V.N. Anaerobic digestion of biomass for methane production: A review. Biomass and Bioenergy 1997;13(1-2) 83-114.

[5]   Willke Th, Vorlop K.-D. Industrial bioconversion of renewable resources as an alternative to conventional chemistry. Applied Microbiology and Biotechnology 2004;66(2)131-142.

[6]   Nishio N, Nakashimada Y. Recent Development of Anaerobic Digestion Processes for Energy Recovery from Wastes. Journal of Bioscience and Bioengineering 2007;103(2) 105-112.

[7]   Yang Y, Tsukahara K, Sawayama S. Biodegradation and methane production from glycerol-containing synthetic wastes with fixed-bed bioreactor under mesophilic and thermophilic anaerobic conditions. Process Biochemistry 2008;43(4) 362-367.

[8]   Bagi Z, Ács N, Bálint B, Horváth L, Dobó K, Perei K R, Rákhely G, Kovács K.L. Biotechnological intensification of biogas production. Applied Microbiology and Biotechnology 2007;76(2) 473-482.

[9]   Johnson D T, Taconi K A. The Glycerin Glut: Options for the Value-Added Conversion of Crude Glycerol Resulting from Biodiesel Production. Environmental Progress 2007;26(4) 338-348.

[10] Temudo M F, Poldermans R, Kleerebezem R, Van Loosdrecht M C M. Glycerol Fermentation by (Open) Mixed Cultures: A Chemostat. Study. Biotechnology and Bioengineering 2008;100(6,15)1088-1098.

[11] da Silva G P, Mack M, Contiero J. Glycerol: A promising and abundant carbon source for industrial microbiology. Biotechnology Advances 2009;27(1)30-39.

[12] Chatzifragkou A, Makri A, Belka A, Bellou S, Mavrou M, Mastoridou M, Mystrioti P, Onjaro G, Aggelis G, Papanikolaou S. Biotechnological conversions of biodiesel derived waste glycerol by yeast and fungal species. Energy 2011;36(2) 1097-1108.

[13] Pagliaro M, Ciriminna R, Kimura H, Rossi M, Della Pina C. From Glycerol to Value-Added Products. Angewandte Chemie - International Edition 2007;46(24)4434-4440.

[14] Fernando S, Adhikari S, Kota K, Bandi R. Glycerol based automotive fuels from future biorefineries. Fuel 2007;86(17-18)2806-2809.

[15] Sosnowski P, Wieczorek A, Ledakowicz S. Anaerobic co-digestion of sewage sludge and organic fraction of municipal solid wastes. Advances in Environmental Research 2003;7(3) 609–616.

[16] Ministry of the Environment, Government of Japan. http://www.env.go.jp/policy/chie-no-wa/download/0502/0502d-2.pdf

[17] Esteves I, Lopes M, Nunes P, Mota J. Adsorption of natural gas and biogas components on activated carbon. Separation and Purification Technology 2008:62 281-296

[18] Carter J W, Wyszynski M L. The pressure swing adsorption drying of compressed air. Chemical Engineering Science 1983;38(7) 1093-1099

[19] Ruthven D M. Principles of Adsorption and Adsorption Processes (10th Ed.). New York: John Wiley; 1984.

[20] Itoga K: Abe Y, Tachimoto H, editor. Kasseitan no Ohyohgijutsu. Tokyo: Technosystem Co, Ltd; 2000. P226

[21] Patterson T, Esteves S, Dinsdale R, Guwy A. An evaluation of the policy and techno-economic factor affecting the potential for biogas upgrading for transport fuel use in the UK. Energy Policy 2011:39 1806-1816

Approaches for the Detection of Toxic Compounds in Castor and Physic Nut Seeds and Cakes

Keysson Vieira Fernandes and
Olga Lima Tavares Machado

Additional information is available at the end of the chapter

1. Introduction

The worldwide search for new fuel sources has grown during the last decades due to two main factors: the global concern about environmental issues and the high price of petroleum. Biodiesel is a type of biofuel that is already used in many countries, and its usage will most likely increase over the next few years. Biodiesel can be produced using different technologies and raw materials, such as vegetable oils, animal fats and microalgae oil. However, despite the wide range of oil sources for biodiesel production, vegetable oils are primarily used for this purpose. The choice of oilseed to be planted for biodiesel production depends on many factors, including the regional climate and soil conditions. The biodiesel industries in the US primarily use soybean oil, whereas in Europe, rapeseed is primarily used for biodiesel production. In tropical countries, biodiesel is produced from plants that grow in these tropical areas, such as palm, physic nut and castor bean.

In addition to biodiesel production using vegetable oils, the by-products generated at different steps during the production process have garnered increasing attention. Some of these by-products are generated in large amounts, making it both economically necessary and interesting to find a use for them. Currently, the residual cake, also known as the seed cake or press cake, has been shown to be a noteworthy by-product. The seed cake consists of the organic waste obtained during the oil extraction process by the pressing of seeds. Large amounts of residual cakes are generated during the oil extraction process. For example, for each ton of castor bean pressed for oil, a half-ton of cake is produced [1]. The residual cake can be used as fertiliser because of the macro- and microelements composition. Moreover, the protein content makes it useful as a component of animal feed.

Several countries from South and Central America and Asia are attempting to use new oil-seed sources for biodiesel production. Two of the oilseeds that are expected to be used for this purpose are the castor bean (*Ricinus communis*) and physic nut (*Jatropha curcas*). The oil properties of these seeds are well known, and many processes have been developed to produce biodiesel from these seeds. However, the large amount of residual cakes that are produced during the biodiesel production process and how to dispose of or use these cakes remain a problem. Both the castor cakes (castor bean) and Jatropha cakes (physic nut) have great potential for use as fertilisers. Castor cakes are rich in macroelements, including N, P, K, Na, Mg and S, and were shown to supplement the nutritional requirements of plants, reduce the soil acidity by increasing the pH, increase the carbon content, reduce the presence of nematodes and promote overall soil health [2]. Jatropha cakes are already used as green manure, also because of the N, P and K content [3]. It is expected that the castor and Jatropha cakes can be used as animal feedstock. These oilseed cakes are high in protein; therefore, their use as an animal food supplement is highly desirable. However, the presence of toxic substances in the seeds of *R. communis* and *J. curcas* restrict the use of the residual cakes as feedstock. Many detoxification processes have been described to render castor and Jatropha cakes edible. However, there is currently no recognized standard and safe methodology that could be used in the industry. Most of the detoxification processes developed have some negative aspects, such as high prices that are limiting for use on an industrial scale or the validation method. This second problem is the most difficult to solve because it is necessary that the detoxified cakes be safe to use as animal feedstock. A flawed method to detect toxins in the cakes could be very dangerous because a non-detoxified residual cake could be used to feed animals and may lead to death. In addition to toxic components, it was shown that allergenic proteins are also present in the seeds of *R. communis*[4] and *J. curcas*[5], and many methods for the detoxification of residual cakes have been shown to efficiently eliminate the toxins but not the allergens. For example, during the 1960s, a detoxified castor cake was commercialised in Brazil as *Lex Proteic* [6]; however, despite the absence of toxins, the allergens remained present in the castor cake. In this chapter, different methods to detect toxins from *R. communis* and *J. curcas* will be described.

2. *Ricinus communis* toxins

Castor bean seeds have long been known for their toxicity. They are the source of the most potent phytotoxin known, the protein ricin. Moreover, the toxic alkaloid, ricinin, is also found in the castor bean; however, this compound is different from ricin in that it is not as toxic and can easily be removed from the castor cake.

2.1. Ricin

The toxin, ricin, has been known since ancient times because of its use in criminal practices. According to Olsnes [7], in 1887, Dixon had hypothesised that the *R. communis* toxin was a protein, and Kobert confirmed this hypothesis in 1913.

Ricin is a type 2 ribosome-inactivating protein (RIP) that is found exclusively in the endo-sperm of castor bean seeds. As a type 2 RIP, ricin is a dimeric protein comprised of an A chain (32 kDa) and a B chain (34 kDa) linked by a disulfide bond [8]. The ricin A chain (RTA) is responsible for the enzymatic activity of the protein. This N-glycosidase enzymatic activity removes a specific adenine, depurination, (A_{4324}) residue from a region of rRNA known as the α-sarcin/ricin loop (SRL) (Figure 1).

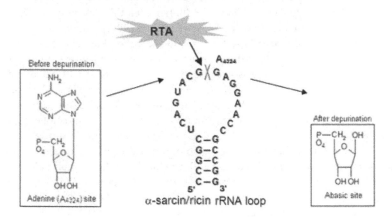

Figure 1. The α-sarcin/ricin loop and the point of depurination by RTA N-glycosidase activity. The A_{4324} site before dep-urination by RTA is shown on the left where the intact nucleotide is present. On the right an after depurination abasic site (without the adenine) is shown.

The absence of this adenine residue inhibits binding of the elongation factor, thereby stop-ping protein synthesis [9]. The B chain (RTB) is a lectin that binds to glycoproteins and gly-colipids on the cell surface and cytosol and mediates the internalisation and intracellular translocation of the toxin [10,11].

The ricin toxin is very efficient and a single molecule may inactivate 2,000 ribosomes per mi-nute [12]. Because ricin can be used as a bioterrorism agent [13], many assays to detect ricin have been described. Some of these assays are highly accurate and can detect very low con-centrations of the toxin. However, there is no standard methodology to use as a quality con-trol for castor cake detoxification processes. Many methodologies to eliminate ricin toxicity from castor cakes have been described, and there are several promising processes when eco-nomic aspects are considered [14]. Therefore, to use castor cakes as animal feedstock, effi-cient methods to detect ricin toxicity after the detoxification process are needed to ensure quality control and safety before the material can be commercialised.

2.1.1. Detection of ricin

Because ricin can be used as a bioterrorism agent, the search for fast and sensitive detection methods began soon after the first studies describing the mechanism of action of ricin. The earliest proposed detection method was the enzyme-linked immunosorbent assay (ELISA)

[15]. In this assay, rabbit anti-ricin antibodies (reduced IgG and Fab' fragments) conjugated with β-D-galactosidase was used. Using the rabbit anti-ricin Fab'-β-D-galactosidase complex, it was possible to detect as little as 4 ng/mL of ricin with the sandwich ELISA technique. However, less sensitivity was observed when this method was utilised for determining the amount of ricin added to rabbit body fluids. In this case, the lowest concentration of ricin that could be assayed was 40 ng/mL. During the next two years, new methods based on radioimmunoassays were proposed [16, 17]. These radioimmunoassays were very sensitive and could detect 50-100 pg RTA and 500 pg RTB; however, the sensitivity was reduced to intact ricin. The matrix used for these assays consisted of 0.1% sodium azide and 0.1% bovine serum albumin (BSA) in 0.05 M sodium phosphate buffer. Limitations of these assays include the difficulties in handling radioisotopes and the long incubation period. Therefore, despite the high sensitivity of these assays, the drawbacks associated with radioimmunoassays make them less preferable than ELISA. Poli et al. [18] developed an enhanced colorimetric and chemiluminescent ELISA to detect ricin in biological fluids. This assay utilised an affinity-purified goat polyclonal antibody (pAb) to adsorb ricin from the solution. The same pAb was then used to form a sandwich, and avidin-linked alkaline phosphatase was used for colour development. Enhancement of the colourimetric assay was obtained because of the increased biotinylated antibody content and a reduction in the dilution ratio of the avidin-linked alkaline phosphatase. This assay could detect 100 pg/mL ricin in phosphate-buffered saline (PBS), human urine and human serum. This sandwich assay could also be used with a chemiluminescence detection reagent; however, the quantitation was limited to a range of 0.1–1 ng/mL and was subject to greater variability compared to the colourimetric assay. An ELISA using monoclonal antibodies (mAb) was performed to detect ricin in biological fluids [19]. This method was also based on the sandwich format using an anti-ricin B chain mAb to adsorb ricin from the solution and an anti-ricin A chain mAb conjugated to peroxidase as the second antibody that is then used to form a sandwich. The peroxidase allows for colour development and measurement of optical density at 450 nm. The sensitivity of this assay is 5 ng/mL and is lower than the sensitivity reported for the amplified and chemiluminescent immunoassays [18]. The ELISA is still used to detect ricin, and a commercial ELISA kit specific for ricin detection can be obtained [20]. However, ELISA has several disadvantages that prevent it from being the best method to detect of ricin. ELISAs consume too much time because of the washing steps involved and they also have limited throughput. ELISAs may also underestimate the actual ricin content in situations where antigen concentrations are high (hook effect) and specialised personnel are also required to perform the ELISAs.

To reduce the time necessary to assay for ricin, a method based on a fiber-optic sensor was developed [21, 22] and optimised [23]. A sandwich immunoassay scheme was used in which an anti-ricin IgG was immobilised onto the surface of an optical fiber. The limits of detection for ricin, as detected by laser-induced fluorescence, in a buffer solution and river water were 100 pg/mL and 1 ng/mL, respectively. The complete assay can be performed in 20 minutes.

The first immunochromatography assay to detect ricin was performed using antibody anti A-Chain mAb with two distinct specificities. An anti-RTB mAb (1G7) was immobilised to a defined detection zone on a porous nitrocellulose membrane, whereas an anti-RTA mAb (5E11) was conjugated to colloidal gold particles that worked as the detection agent [24]. The ricin-containing mixture was added to the membrane and allowed to react with the mAb 5E11-coated particles. This mixture moved across the porous membrane by capillary action until it reach the extremity containing the anti-RTB mAbs, which bound to the particles of ricin that were attached to the gold-labelled anti-RTA mAbs. The detection limit of this assay was 50 ng/mL ricin in phosphate-buffered saline (PBS). This sensitivity could be enhanced further to 100 pg/mL with the use of a silver enhancer. The advantages of these gold particles were their superior mobility, decreased aggregation and commercial availability. An immunochromatography assay was also used to shown differences in ricin content among different castor bean cultivars [25]. All the ricin isoforms were detected in the range of 1 to 2.5 ng/ mL in buffer.

In addition to using a better antibody for improved sensitivity, there was also a development regarding the technology of the solid phase surface of the immunoassay. The conventional microplate was exchanged for magnetic micro beads. Immunomagnetic (IM) assays to detect ricin were first used by Gatto-Menking et al. [26]. They used immunomagnetic electrochemiluminescence (IM-ECL) to detect ricin and other toxic agents, such as botulinus A, cholera β subunit, ricin and staphylococcal enterotoxoid B. Antibody-conjugated magnetic micro beads were used to capture the target toxins and ruthenium trisbipyridal chelate-labelled antibodies were used as the reporter. High sensitivity levels were obtained for all the tested toxins. All IM-ELC assays could be performed in a maximum combined incubation and assay time of approximately 40 minutes, and the sensitivity to ricin was 5 pg/mL. Some years later, an enhanced ECL assay had a detection limit of 0.5 pg/mL for ricin in PBS [27]. The same study demonstrated the detection of ricin by fluorogenic-chemiluninescence (FCL), and the sensitivity was 1 ng/mL. Advantages of these micro beads were due to their large surface area (Figure 2) that leads to enhanced sensitivity, to free moving microspheres coated with antibody that accelerates the reaction rates and reduces the assay time, and to easy detection using a simple magnetic field. Both the FCL and ECL had similar formats, except that the FCL used alkaline phosphatase as the label and detected the ricin through the measurement of fluorescence, whereas the ECL used ruthenium-trisbipyridal as the label and detected the ricin through photoemission. For a magnetoelastic surface sensor instead of microspheres, the detection technology was a sandwich immunoassay on the sensor surface. Biocatalytic precipitation was then used to cause a change in mass, which resulted in a change in the resonance frequency that allowed for quantitation of ricin at a detection limit of 5 ng/mL in aqueous media, such as water, blood or serum [28]. This magnetoelastic sensor had a sensitivity that was comparable to the ELISA; however, this assay had a much lower cost, was disposable and had a relatively quick analysis time.

The search for an assay to detect several toxins simultaneously led to the use of array systems. Three different toxins, ricin, SEB and Yersinia pestistoxin, were detected using a planar array immunosensor equipped with a charge-coupled device (CCD) [29]. This was a disposable and simple sensor array coated with different antibodies that were detected through the CCD. This planar array platform gave a detection limit of 25 ng/mL ricin, 5 ng/mL SEB and 15 ng/mL *Y. pestis*, based on a goat anti-ricin antibody in PBS containing 0.05% (v/v) Tween-20. This detection method allowed for multiple sample analysis using a minimum amount of sample and simultaneous analysis that was inclusive of the controls. An antibody microarray biosensor for the rapid detection of both protein and bacterial analytes under flow conditions was developed using a micrometer-sized spot [30]. Using a non-contact microarray printer, biotinylated capture antibodies were immobilised at discrete locations on the surface of an avidin-coated glass microscope slide. The slide was fitted with a six-channel flow module that conducted analyte-containing solutions over the array of capture antibody microspots. Detection of the bound analyte was subsequently achieved using fluorescent tracer antibodies. The pattern of fluorescent complexes was interrogated using a scanning confocal microscope equipped with a 635-nm laser. The assays were completed in 15 minutes, and ricin detection was demonstrated at levels of 10 ng/mL.The detection limits for the other analytes were also relatively low. These assays were very fast compared to the previously published methods for measuring antibody-antigen interactions using microarrays (minutes versus hours). In addition, whereas other antibody microarray assays can detect specific proteins present in complex mixtures, this method could detect proteins and bacteria simultaneously. Recent improvements in the microarrays to detect ricin and other biological agents have been described. A method that used a bioanalytical platform that combined the specificity of covalently immobilised capture probes with dedicated instrumentation and immuno-based microarray analytics was able to detect ricin at 0.5 ng/mL in PBS and 1-5 ng/mL in milk [31]. However, despite the high sensitivity compared with the other array methods, this assay took approximately 90 minutes.

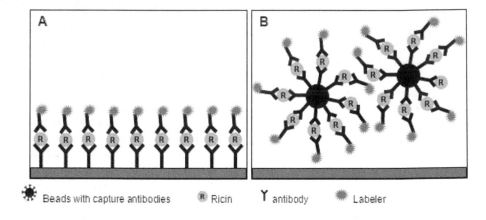

Figure 2. Comparative schematic of two immunoassays used to detect ricin. A) Sandwich ELISA. B) Microbeads immunoassay. The recorder antibody can be linked to different labeler molecules, as Ruthenium, alkaline phosphatase or horsehadish peroxidase.

Sano et al. [32] developed a method to detect antigens that combined the specificity of immunological analysis with the exponential amplification of PCR. This immuno-polymerase chain reaction (IPCR) was an interesting method to monitor the presence of ricin in samples [33]. A schematic representation of this method is shown in Figure 3.

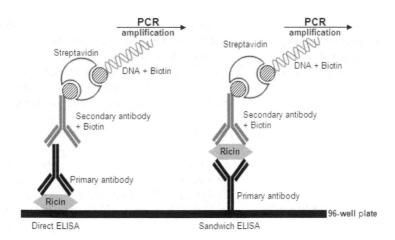

Figure 3. Schematic representation of the IPCR detection of ricin using both direct and sandwich ELISA to capture and report the toxin. The biotin-streptavidin interaction plays the bridge role between secondary antibody and the reporter DNA, which is amplified by PCR.

Ricin was dissolved at different concentrations in PBS, and detection was performed revealing a detection limit of 10 fg/mL. The assay was then performed with ricin dissolved in human serum revealing a detection limit of 0.5 fg/mL. The method has also been used for post-intoxication evaluation of the biological half-life of ricin. IPCR analysis of sera from mice fed ricin showed that the toxin was rapidly sequestered from the sera (30 minutes) with a half-life ($t_{1/2}{}^{\alpha}$) of 4 minutes [34]. The time required to complete the entire IPCR process is 9 hours. Compared with conventional immunological methods, IPCR requires a greater amount of time because of the PCR itself and the post-PCR analysis. Moreover, the use of more expensive reagents and the increased reagent consumption make this technique less attractive than conventional immunological methods. However, these limitations are counterbalanced by greater sensitivity (8 million times greater than conventional ELISA), enabling a broader range of applications.

In recent years, highly sophisticated mass-spectrometry (MS)-based methods for the detection and quantification of ricin have been developed. It was shown that ricin could be unequivocally identified by liquid chromatography-electrospray (LC-ES) MS/MS experiments with reduced, cysteine-derivatised, trypsin-digested material [35]. It was also shown that MALDI-MS could be used to detect intact ricin and to screen samples for ricin peptides. The amount of crude sample required was a few milligrams containing less than 5% ricin. According to the authors, the selection of a few marker peptides from the A and B chains can be used as a method to improve the sensitivity and efficiency of this method. A method combining immunocapture and analysis by matrix-assisted laser desorption/ionisation time-of-flight mass spectrometry (MALDI-TOF MS) for ricin detection was also described [36]. Ricin samples were applied to magnetic spheres coated with a monoclonal anti-B-chain antibody. After acidic elution, tryptic peptides of the A and B chains were obtained by accelerated digestion with trypsin in the presence of acetonitrile. Three of the 20 peptides obtained were used for ricin detection by MALDI-TOF MS. This assay had a limit of detection estimated at 50 ng/mL, and the result could be obtained in approximately 5 hours. These results are not as exciting compared to other more sensitive and faster methodologies; however, an interesting feature is that MS detection provides increased specificity because of the simultaneous monitoring of several characteristic ricin-specific peptides. Furthermore, the possible miniaturisation of MALDI-TOF technology suggests that the assay could be adapted for use with a portable mass spectrometer. A recent study described the combination of a multiplex-immunoaffinity purification approach followed by MALDI-based detection for the simultaneous identification of different toxins, including ricin [37]. Selected antibodies against each toxic agent allowed for the specific and simultaneous capture of these toxins. The toxins were subsequently identified by MALDI-TOF MS following a tryptic digest, and after an assay time of 8 hours, the ricin could be detected at a minimum of 200 ng/mL. The time requirement and detection limit were not satisfactory for this assay; however, ricin could be detected in complex matrices, such as milk and juice.

Aptamers are artificial nucleic acid ligands that can be generated against amino acids, drugs, proteins and other molecules. They are isolated from complex libraries of synthetic nucleic acids by an iterative process of adsorption, recovery and reamplification. Because of their high thermostability when compared with antibodies, aptamers have potential applications in analytical devices, including biosensors, and as therapeutic agents [38]. Assays for protein identification and quantitation were developed and applied to ricin detection [39, 40]. A multiplex aptamer microarray was generated by printing an anti-ricin RNA aptamer onto either streptavidin (SA)- or neutravidin (NA)-coated glass slides. The limit of detection in a sandwich assay format after optimisation studies was 15 ng/mL in PBS. This assay was also used to detect other proteins and showed satisfactory results. Capillary electrophoresis (CE) has been shown to be a viable alternative to traditional immunoassays when coupled with laser-induced fluorescence detection. Haes et al. [41] demonstrated that capillary electrophoresis could be used to detect ricin by monitoring its interaction with a fluorescently tagged aptamer under non-equilibrium conditions. The quantitative response revealed a detection limit as low as 14 ng/mL. This study also revealed that the presence of nucleases in the sample leads to a slight decrease in the ability of the aptamer to detect ricin; however, it is still possible to detect the toxin at very low concentrations. This assay can be performed in less than 10 minutes, consumes minimum quantities of material, and generates a low amount of waste.

Liquid-crystal (LC) based sensors that can be used as rapid and effective detection technologies have attracted a significant amount of attention in recent years [42], and their utility regarding ricin detection has previously been demonstrated [43]. This method relied on the use of LCs 5CB to amplify and report the presence of ricin captured by an affinity ligand.

One merit of this approach is that the ricin can be imaged on chemically functionalised surfaces and transduced into an optical signal. The optical signal caused by the orientational transition of the LCs could easily be identified with polarised light microscopy. However, despite the success of the LC-based sensor, which did not use complex instrumentations and did not involve any labelling steps, the limit of detection of 10μg/mL was not as good compared to other methods. Similar to other assays, this interesting technology must be improved to become among the most sensitive methods for ricin detection.

Despite the many methods to detect the presence of ricin, the detection of the toxin in castor cakes subjected to detoxification is not performed in a standard manner. Anandan et al. [44] used different physical and chemical treatments to detoxify castor cakes, and the ricin content was determined based on electrophoretic analysis. They reported that ricin bands did not appear in SDS-PAGE samples of autoclaved (15 psi, 60 minutes) and lime treated (40 g/kg) castor cakes. Solid-state fermentation by *Penicillium simplicissimum* also reduced the ricin content when fermented castor waste samples, which were not the cake but an extremely alkaline waste, were evaluated by electrophoresis [45]. However, this detection method has many disadvantages compared to the described techniques. The first disadvantage is the low sensitivity of the method. A lower ricin concentration that remains lethal cannot be detected; therefore, if electrophoresis is used as the detection method, another more sensitive assay needs to be performed to validate the detoxification process. Another problem is related to the long assay time and specialised personnel required to perform these analyses and the necessity of performing a Western blot assay to confirm the identity of ricin.

The greatest problem that affects not only electrophoresis, but also all the ricin detection methods described in this chapter, is the inability to detect the biological activity of the toxin. Each proposed assay can detect the presence of ricin at minimal concentrations and many of these are able to do so in a very sensitive and specific way; however, they cannot determine whether the toxin is biologically active. To validate the castor cake detoxification processes, it is important to be able to detect the biological activity of ricin. This is because some of the described toxin inactivation processes can be related to modifications in the active site of the enzyme, and although ricin may be present in processed cake, it may be not active and the product would be safe to use in animal feed.

2.1.2. Detection of ricin biological activity

The first method of detecting ricin activity was based on measuring the inhibition of protein synthesis in a rabbit reticulocyte cell-free system mediated by toxic tryptic peptides from ricin [46]. The method was justified because of the long period of time required to observe intoxication symptoms in animals. It was reported that similar to the native protein, toxic ricin peptides could inhibit protein synthesis in a cell-free system. This information reinforces the necessity for assaying ricin biological activity after subjecting the castor cake to detoxification processes.

The ability of the RIPs in inhibit protein synthesis can be monitored with *in vitro* translation assays using the rabbit reticulocyte lysate system [47, 48]. One disadvantage of these assays is the use of a multistep procedure to determine the RIP activity by measuring the incorporation of radioactive amino acids after the addition of mRNA or polysomes to the system. Therefore, an *in vitro* transcription/translation single-step assay utilising the luciferase bioluminescence detection system was described to characterise mistletoe lectin I (ML-I) and ricin [48]. The *in vitro* translation assay couples the following reactions into one step: (1) DNA consisting of a coding sequence is transcribed into messenger RNA; and (2) RNA is then translated into proteins in a cell lysate (product of burst cells) that provides ribosomes and other necessary components. When the translated protein is luciferase, the fluorescence acts as a protein synthesis indicator, and the absence of fluorescence indicates that protein synthesis was inhibited (Figure 4). The inhibition of luciferase synthesis by ricin was achieved when the toxin was used at a minimum concentration of 30.2 pM (~800 pg/mL). The RIP specificity of this assay was proved using formycin 5'-monophosphate (FMP) as a specific inhibitor of RIP activity. The limit of detection is comparable to those obtained with other methodologies, and the assay also showed the toxic activity of ricin.

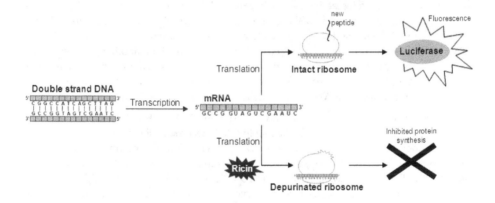

Figure 4. Translation and protein synthesis inhibition by ricin. The assays based on this activity detect the presence of an specific reporter protein. In the presence of the protein can not be synthesized. The luciferase is the best described example for this kind of assay.

The inhibition of protein synthesis was also the target of a method to detect ricin in a "well-in-well" device [49]. The miniaturised system presented a mechanism to supply nutrients continuously and remove by-products, leading to higher protein expression yields and larger detection signals. This method showed a detection limit of 0.3 ng/mL ricin. The nested-well device was also used for measuring the toxicity of ricin after physical or chemical treatment. The good results obtained with inactivated ricin make this method a good choice for use in castor cake detoxification processes.

The N-glycosidase activity removes an adenine residue from the α-sarcin/ricin loop of rRNA. The removed adenine can be used as a positive indicator of biologically active ricin. The most common method for quantifying free adenine in a variety of applications is the detection of fluorescent-derivatised adenine by HPLC [50]. To detect ricin activity based on rRNAdepurination, a high-throughput, enzyme-based colorimetric adenine quantification assay was developed [51]. The key step of this assay is the conversion of adenine to AMP and concurrent release of pyrophosphate from PRPP. Pyrophosphate is then cleaved to phosphate by inorganic pyrophosphatase. To enhance the signal, the AMP formed is converted by 5'-nucleotidase to adenosine and inorganic phosphate, finally resulting in three phosphates for each adenine. Inorganic phosphate was quantified by a modified procedure with a commercially available kit. All four enzyme reactions of the assay, including colour development, occur simultaneously in approximately 15 minutes inside the same reaction tube, and the rate of adenine released by the commercially obtained RTA was determined to be 43 pmol adenine/pmol RTA per hour.

Recently, several methods using electrochemiluminescence (ECL) to detect ricin activity were also developed [52, 53]. First, a deadenylation assay using paramagnetic beads could detect ricin in crude extracts [52, 54]. Synthetic biotinylated RNA substrates were cleaved by the combined actions of the ricin holotoxin and a chemical agent, N,N'-dimethylethylenediamine. The annealing of the product with a ruthenylatedoligodeoxynucleotide resulted in the capture of ruthenium chelate onto magnetic beads, enabling the electrochemiluminescence (ECL)-based detection of RNA N-glycosidase activities of toxins. Compared to ECL immunoassays [26], the ECL activity assay presented lower sensitivity, reaching a detection limit of 100 pg/mL. The disadvantage of the ECL immunoassay compared to the ECL activity assay is that the antibodies recognise surface features of the proteins (epitopes) that may be unrelated to any enzymatic activity or other mechanism of toxicity. Therefore, it may be possible for inactive protein toxins to cause positive signals in these immunoassays resulting in an over-estimation of the threat. The plate-based assay unlike the bead-based assay, included wash steps that enabled the removal of food particles, thereby maximising the matrix effects and improving the limits of detection. The limits of detection for ricin in apple juice, vegetable juice, and citrate buffer using the bead-based assay were 0.4, 1, and 0.1 μg/mL, respectively. By contrast, the limits of detection for ricin using the plate-based assay were 0.04, 0.1, and 0.04 μg/mL in apple juice, vegetable juice, and citrate buffer, respectively. These data suggest that the plate-based assay is the best method for detecting ricin activity by ECL.

The ricin detection methods based on adenine liberation and direct infusion electron spray ionisation mass spectrometry have been shown to provide rapid, selective, and sensitive detection of various peptides and small nucleic acids, and these methods should provide a sensitive method for the real-time analysis of RIP enzymatic activity by monitoring adenine release. Therefore, high-performance liquid chromatography (HPLC) and selected ion monitoring mass spectrometry (MS) were used to develop a quantitative assay for adenine release from a synthetic RNA substrate by the ricin A chain [55]. The sensitivity of this MS assay made it possible to measure RIP activity at approximately 0.6- to 600 ng/mL. A more specific assay to detect ricin by MS was developed by Becher et al. [56] in which they used an anti-B chain mAb immobilised on magnetic beads to capture the toxin. Ricin toxicity was measured through quantification of the free adenine by HPLC-MS. The immunoaffinity step combined with enzymatic activity detection led to a specific assay for the entire functional ricin protein with a lower limit of detection of 100 pg/mL.

When mass spectrometry was used to detect ricin activity, a combination of three techniques, all performed on the same sample, provided a sensitive and selective analysis of ricin isolated from a food or clinical sample and measured the activity of the toxin [57]. First, ricin was isolated from abundant proteins in a food or clinical sample, such as milk, apple juice, serum or saliva through immunoaffinity capture on antibody-coated beads. Second, the activity of ricin was examined through interaction of the toxin with a DNA substrate that simulated the *in vivo* target of the toxin. The DNA substrate was analysed by MALDI-TOF MS, allowing for sensitive and selective measurements of the depurination of the DNA substrate. Finally, in the third step, the ricin was subjected to tryptic digestion, and the resulting tryptic fragments were analysed by liquid chromatography tandem mass spectrometry (LC-MS/MS), allowing for direct examination of the composition of the ricin protein based on the molecular weight change caused by the depurination activity. The limit of detection was approximately 300 ng/mL.

The mass spectrometry based methods for detecting ricin activity through monitoring adenine liberation have some disadvantages that make them not suitable for use in the validation of the detoxification processes of the castor cakes. These disadvantages include complications regarding the handling of mass spectrometers and the interpretation of results that requires highly specialised personnel. Another problem is that adenine liberation may not be the most efficient method to detect biologically active ricin because depurination activity is not a unique mechanism involved in ricin toxicity. It was previously shown that non-cytotoxic RTA mutants could depurinate ribosomes in yeast cells without the occurrence of cell death and apoptosis signals [58].

Toxicology assays to detect ricin based on the activity against animals could be the best way to evaluate the efficiency of castor cake detoxification processes because of the desire to use this by-product as animal feedstock. However, despite the ethical questions surrounding the use of *in vivo* models, there are also economic and infrastructure problems. Housing live animals to evaluate toxic activity requires physical space and maintenance. Therefore, an *in vitro* assay based on the cytotoxicity against Jurkat clone E6-1 cells was developed to detect ricin in different beverages, such as orange juice, coffee and soda, and food matrices, such as milk, milk baby formula and soy baby formula [59]. After incubating the cells in a 96-well plate with ricin, the culture was maintained overnight at 37ºC and 5% CO_2. Aliquots of each treated well were collected and assayed for lactate dehydrogenase (LDH) activity with a colorimetric assay. LDH was released from the cytosol upon cell damage and was positively correlated with cell death. Ricin was detected in each assayed matrix with a sensitivity of 10-100 pg/mL. It was also shown that ricin cytotoxicity could be inhibited by the administration of an anti-ricin neutralising antibody that works as a qualitative mechanism. Other cell culture assays were also recently developed. Sehgal et al. [60] used Vero cells (*Chlorocebus sabaeus* kidney cells) to evaluate the cytotoxicity of different ricin isoforms. They showed that the isoforms R-I, R-II and R-III were detected at a minimum concentration of 20 mg/mL, 10 ng/mL and 2 ng/mL, respectively. Subsequently, they showed that the cytotoxicity of the three isoforms is time dependent and that the R-III isoform is more glycosylated than the other two isoforms [61].

The possibility of using cell culture models to evaluate ricin toxicity by colorimetric assays, such as the LDH assay, seem to be a good idea for use as a biological test to determine the efficiency of the castor bean cake detoxification process. It was reported that solid-state fermentation (SSF) reduced the ricin levels in castor bean alkaline waste from Petrobras (the national petroleum company of Brazil) during the biodiesel production process [45, 62]. This was determined by molecular exclusion chromatography and electrophoresis. To verify the biological activity of ricin after SSF at different time intervals, an *in vitro* assay using the Vero cell line was performed [63]. Using this methodology, it was verified that after 24 and 48 hours of fermentation, the cell culture showed slight growth inhibition. The waste was completely detoxified after only 72 hours of fungal growth. The cell incubation period with the protein extract from the fermented waste was 24 hours, and cell death was determined by cell counting with an optical microscope and measurement of LDH activity using a colorimetric assay (Figure 5).

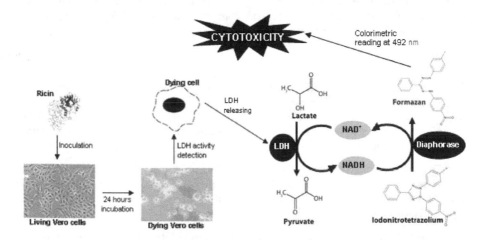

Figure 5. Cytotoxicity assay using Vero cells. The cell monolayer in a 24-well plate was incubated with ricin for 24 hours. An aliquot from each well was removed and mixed with the assay solution for LDH activity determination. The formazan formed a salt that caused the solution to turn red. The cytotoxicity is measured based on the intensity of this colouration.

When the cell counting and LDH assays were compared to determine the cytotoxicity of ricin against Vero cells, it was reported that both methods are efficient and detected ricin at a minimum concentration of 10 ng/mL [64]. After adjusting the method to detect the purified protein, they used the Vero cell cytotoxicity assay to evaluate the following two castor cake detoxification processes: SSF using *Aspergillus niger* and treatment with calcium compounds. The results with the Vero cells showed that both treatments were efficient in eliminating ricin toxicity from the castor cake.

3. *Jatropha curcas* toxins

Two main toxic components are present in the physic nut plant, the ribosome-inactivating protein, curcin, and phorbol esters. Among these toxins, the phorbol esters are the most dangerous toxic components in *J. curcas* and limit the use of Jatropha cake in animal feed.

3.1. Curcin

Curcin (28.2 kDa) is a type 1 RIP that is found in *Jatropha curcas* seeds [65] and leaves [66]. Curcin is different from ricin in that it is a monomeric protein with N-glycosidase activity but lacks a lectin chain [67]. Therefore, this protein is much less cytotoxic than ricin and other type 2 RIPs because it cannot enter cells by binding to sugar residues. Despite the fact that curcin is less toxic than phorbol esters, it has been reported to be toxic to some animals, including sheep, goats, chickens and calves and also to humans [68-72]. Because of the low toxicity of curcin, there are not many detection methods specifically for this toxin. The most common detection methods are the inhibition of translation in rabbit reticulocyte lysates and the measurement of N-glycosidase activity [67]. Although there are few publications describing the different methods to detect curcin, many of the assay methods for ricin could be applied to other RIPs, including curcin.

3.2. Phorbol esters

Phorbol esters (PE) are polycyclic compounds in which two hydroxyl groups in neighbouring carbons are esterified to fatty acids, and these substances are present in many different plants, including *J. curcas* [73]. The PE molecules are dependent on a tetracyclic diterpene carbonic structure termed tigliane. The different hydroxylation points of tigliane determine the different varieties of PE and their toxicity [74].

The PEs and their different derivatives are known for their tumour induction activity. They activate protein kinase C (PKC), which plays a critical role in signal transduction pathways and regulates cell proliferation [75]. By contrast, it was reported that some types of PEs could induce apoptosis [76].

Several detoxification processes used to eliminate PEs from Jatropha cake have been previously described [14], and some of the existing detection methods were used to confirm the effectiveness of these processes.

3.2.1. Detection of phorbol esters

Many of the phorbol ester detection methods are related to using Jatropha cake as animal feedstock, which are different from ricin containing cakes that can be used as bioterrorism agents. Therefore, there are few techniques for PE detection compared to ricin detection methods.

The determination of irritant activity caused by phorbol esters was first demonstrated by Adolf et al. [77]. The irritant activity of PE isolated from different Jatropha species was assayed in rat ears, and the irritant dose 50 for *J. curcas* PEs was 0.02 µg/ear. More than two decades later, *in vivo* studies of PE toxicity are still performed in rats and mice [78, 79]. Jatropha cake subjected to alkali and heat treatments to reduce the PE level was used to feed rats, and several clinical aspects and the mortality rate were compared with rats fed untreated cake [78]. Using these rodents to detect PE toxicity was effective because even the treated Jatropha cakes with low levels of PE (8.1 mg%) caused rat mortality after 11 days. The acute toxicity of PE was determined in Swiss Hauschika mice by intragastric administration [79]. The LD_5 and LD_{95} were 18.87 and 39.62 mg/kg body mass, respectively. These toxicity assays efficiently detect PE toxicity; however, they are problematic because of maintaining and sacrificing many animals due to the large quantities of residual cake that is generated.

The most commonly used method to detect and quantify PE from *Jatropha curcas* is reverse phase - high-performance liquid chromatography (RP-HPLC). This method was standardised to detect PE in different provenances of *J. curcas*, and it was the first method to identify the absence of PE in seeds from Papantla, Mexico [80, 81]. The protocol established in this study has been optimized [82] to show the presence of PE. The limit of detection of PE by RP-HPLC analysis is approximately 4 µg, as described by Devappa et al. [83]. RP-HPLC detection has been used by many researchers to determine the efficiency of Jatropha cake detoxification processes, including hydrothermal processing techniques, solvent extraction, solvent extraction plus treatment with $NaHCO_3$, ionizing radiation, heating, bio-detoxification and surfactant solution extractions [84-88]. This technique can also be used to identify different PE species present in *J. curcas*, and the difference in PE composition among Jatropha seeds from different regions, cultivars and assessments [83, 89-91]. HPLC was also used to determine the PE content in oil extracted from the seeds [92-94].

Similar to ricin detection methods, the biological activity of phorbol esters must to be assayed to guarantee the efficiency of the Jatropha cake detoxification processes. Because Jatropha cake is used as feedstock, quality control of detoxification processes is often performed using live animals, such as rats [77, 79]), sheep [95], pigs [96] and fish [97, 98]. With a few exceptions, this type of biological activity control is usually preceded by RP-HPLC detection and quantification of PEs. Therefore, it remains necessary to continue using RP-HPLC and sacrificing animals to detect the presence and biological activity of PEs because toxicity evaluation using live animals is not the best method for use on a large scale. Other biological tests have previously been described for assaying PE toxicity, and some of these assays are very sensitive and simple to perform on a large scale.

Earlier reports regarding *J. curcas* have described molluscicidal activity of the seed extracts against *Oncomelania quadrasi* [99] and of the root extracts against *Bulinus truncatus* [100]. However, the most well-established molluscicidal test using snails was described by Liu et al. [101]. They tested several plant extracts, including *J. curcas* phorbol esters in methanol, against three schistosome vector snails: *Oncomelania hupensis*, *Biomphalaria glabrata* and *Bulinus globosus*. The 4-β-phorbol-13-decanoate was the most effective phorbol ester against the snails. It killed both species (LC_{100}) at a concentration of 10 mg/mL. One disadvantage of this method is the requirement of a large volume of the test substances because the assay must to be performed in 100 mL Petri dishes. However, the assay using snails continues to be used and is sometimes combined with HPLC detection and quantification steps. Another species that was tested for PE toxicity was *Physa fontinalis*, which was sensitive to 0.1 mg/L (6.7% mortality) PE-rich extract, and the LC_{100} was reported as 1 mg/mL [93, 102]). The variation in PE sensitivity among the snails may be related to species-specific PE sensitivity and/or different chemical properties of the PEs. In addition to testing for PE activity against host snails, the susceptibility of the parasite *Schistosoma mansoni*was also assayed [103]. This test had the advantage of requiring a small volume of test substance. The PE-rich methanol extract from *J. curcas* crude oil that was obtained by pressing the seeds was able to kill all the cercarie (LC_{100}) at a concentration of 25 mg/mL.

The efficacy of phorbol esters against insects has been shown recently. Termites (*Odonto- termes obesus*) were used as a target to test PE toxicity [104]. Because it was necessary to use HPLC to isolate and quantify PE from *J. curcas* seeds, they tested different concentrations of PE (500-5 mg/mL) over a period of 1 to 72 hours. The LC_{100} was determined after 72 hours of treatment using 5 mg/mL of PE. However, to decrease the assay time, it was necessary to use higher concentrations of PE. To obtain the LC_{100} after 12 hours of treatment, they used 500 mg/mL of PE. Another study using insects was recently performed by Devappa et al. [105]. They tested a PE-enriched fraction (PEEF) against *Spodoptera frugiperda* and the mortality was evaluated 24 hours after treatment with different concentrations of PEEF. A minimum mortality (20%) was reached using 0.5 mg/mL PEEF and a maximum of 80% mortality was observed with 2 mg/mL PEEF. The sensitivity to PEs of both species (*O. obesus* and *S. frugiperda*) is not very different, and this assay showed that PEs can be used as an insecticide and that insects are good models for detecting the toxic activity of PEs.

Some crustaceans are widely used as toxicity indicators in bioassay systems. Phorbol ester toxicity has previously been assayed to *Artemia salina* and *Daphnia magna* [83]. The advantages of using *A. salina* in toxicological assays were demonstrated by Ruebhart et al. [106]. These advantages include wide commercial availability of the cysts, easy storage, maintenance and hatching of the cysts, the assay is cost effective, simple, rapid and sensitive, less test samples are required, the assays can be performed in 96-well microplates and meets the ethical animal treatment guidelines of many countries. The best PE induced mortality rate (72%) was observed using a concentration of 47 mg/mL [83]. Increasing the concentration did not effectively improve the mortality rate because 6000 mg/mL of PE was needed to reach 100% mortality. Different types of PEs were previously tested against *A. salina* [107] and there was variation in the mortality rates to each PE. This reinforces the role of the PE chemical structure and purity with regard to toxicity. The first toxicological assay of PEs from *J. curcas* using *Daphnia magna* showed that these crustaceans are more sensitive to PEs than *A. salina* [83]. The LC_{100} was only 3 mg/mL, and the lowest effective concentration, which induced 26% mortality, was 0.5 mg/mL. Although snails were more sensitive to PEs than crustaceans, the use of *A. salina* and *D. magna* is preferred for assaying a large number of PE samples because the test can be performed in 96-well plates.

Similar to the molluscicidal, insecticidal and antiparasitic activity, PE toxicity against microorganisms was also reported. It was demonstrated that phorbol esters from *Sapium indicum* had antibacterial activity [108]. Six bacteria genera were recently tested for PE toxicity. The maximum concentration of PE-rich extract for each bacterium tested was 537 μg/mL for *Bacillus subtillis*, 250.7 μg/mL for *Pseudomonas putida*, 215 μg/mL for *Proteus mirabilis*, 394 μg/mL for *Staphylococcus aureus*, 215 μg/mL for *Streptococcus pyogenes* and 465.7 μg/mL for *Escherichia coli* [83]. Compared with the other biological assays presented here, the use of bacteria to detect the toxic activity of PEs is not very effective because the sensitivity is much higher than those reported for *D. magna*. The use of PEs as an antibacterial agent was also not as effective compared with the other compounds. The antifungal activity of *J. curcas*PEs extracted from residual cake has previously been tested [83, 109]. The toxicity of the PE-rich extract (from Jatropha cake) against *Fusarium oxysporum*, *Pythiumaphani dermatum*, *Lasiodiplodia theobromae*, *Curvularia lunata*, *Fusariums emitectum*, *Colletotrichum capsici* and *Colletotrichum gloeosporioides* was assayed and the concentrations that inhibited 100% of mycelial growth was 6, 3, 6, 5, 3, 4and 10 mg/L, respectively. Although a high concentration of PEs was required to reach 100% inhibition, they used 500 μg/mL PEsand reported minimum mycelial growth inhibition values for each species [109]. Another PE-rich extract toxicity study using fungi was recently reported [83]. In this study, it was demonstrated that of seven species of fungi, the most sensitive to PE toxicity were *Botrytis cinerea*, *Fusarium oxysporum* and *Fusarium moniliforme* and 100% inhibition was achieved at a concentration of 114.6 μg/mL. The other four species tested, *Aspergillus niger*, *Aspergillus flavus*, *Curvularia lunata* and *Penicillium notatum*, were less susceptible to PE toxicity and 100% inhibition was reached using 143.3 μg/mL. Antimicrobial tests using bacteria and fungi efficiently detect PE toxic activity and could be used for quality control to determine the effectiveness of Jatropha cake detoxification processes.

Because PEs are activators of protein kinase C (PKC), a biochemical assay to detect PEs based on this property was described [110]. In this method, PKC is incubated with Mg-ATP and a synthetic peptide which is labelled with a fluorescent dye. When a PKC activator is present, the active enzyme phosphorylates the peptide. When the reaction mixture is separated by electrophoresis, the phosphorylated peptide becomes negatively charged and migrates to the positive pole. The fluorescently labelled peptide can then be quantified by densitometric analysis. This assay was used by Wink et al. [110] to determine the activity of PEs sequestered by *Pachycoris klugii*. The positive control (12-O-tetradecanoylphorbol-13-acetate) was used at a concentration of ~6μg/mL and indicated that this activity assay is very sensitive. Because of the high sensitivity and availability of commercial PKC activity assay kits, this method could be used for the rapid and efficient detection of PEs in detoxified Jatropha cake.

Although many methods have been described to detect *Jatropha curcas* phorbol esters, these biological tests are not specific to PEs. In contrast to ricin detection assays that can combine biological assays with antibody recognition [59, 63, 64], PEs cannot be tested with this methodology. The best method to test for PEs is to continue using HPLC analysis followed by a biological test. The most well-established biological assay is the assay using snails, which has previously been used as a quality control for Jatropha cake detoxification [83, 103]. Although several of the *in vitro* assays, such as PKC activity and toxicity against microorganisms are more sensitive, they were not used for this purpose, and additional studies are necessary.

4. Conclusion

Currently, several processes to detoxify castor bean and Jatropha cakes have been developed however, it is essential to choose a method that is universally accepted to validate such processes of detoxification. The literature indicates that the method to be used to evaluate the toxicity of castor cake is different from what should be used for jatropha cake.

Among the different methods that can be used to assess the presence of ricin some are more suitable to control attacks bioterrorist. They are sensitive methods that detect the presence of ricin, but need not evaluate the biological activity.

In this review, methods based on Vero cell viability are best suited to validate the processes of castor cake detoxification. Vero cells, epithelial cell line isolated from African green monkey are indicated since these cells maintain cell organelles characteristics and stable structure when in contact with the cake detoxified. Evaluation procedures for Jatropha are still under development. The detection of phorbol esters by reverse phase chromatography, associated with toxicity tests on snails are recommended.

Author details

Keysson Vieira Fernandes and Olga Lima Tavares Machado

Universidade Norte Fluminense – Darcy Ribeiro (UENF), Brazil

References

[1] Santos RF, Barros MAL, Marques FM, Firmino PT, Requião LEG. Análise Econômica. In: Azevedo DM, Lima EF (ed). *O Agronegócio da Mamona no Brasil*. Campina Grande: Embrapa; 2001. p17-35.

[2] Savy Filho A. *Mamona: Tecnologia Agrícola*. Campinas: EMOPI; 2005.

[3] Kumar A, Sharma S. An evaluation of multipurpose oil seed crop for industrial uses (*Jatropha curcas* L.): A review. Industrial Crops and Products 2008; 28(1): 1-10.

[4] Spies JR, Coulson EJ. The chemistry of allergens VIII. Isolation and properties of an active protein-polysaccharidic fraction, CB-1A, from castor bean. *Journal of the American Chemical Society* 1973; 65(9): 1720-1725.

[5] Maciel FM, Laberty MA, Oliveira ND, Felix SP, Soares AMS, Verícimo MA, Machado, OLT. A new 2S albumin from Jatropha curcas L. seeds and assessment of its allergenic properties. *Peptides* 2009; 30(12): 2103-2107.

[6] Perrone JC, Iachan A, Domont GB, Disitzer LV, Castro VRO, Roitman R, Gomes SM. *Contribuição ao estudo da torta de mamona.* Rio de Janeiro: Departamento de Imprensa Nacional; 1966.

[7] Olsnes S. The history of ricin, abrin and related toxins. *Toxicon* 2004; 44(4): 361-370.

[8] Olsnes S, Wesche J, Falsnes PO. Binding, uptake, routing and translocation of toxins with intracellular sites of action. In: Alouf JE, Freer JH (ed). *The Comprehensive Sourcebook of Bacterial Toxins.* London: Academic Press; 1999. p73-93.

[9] Endo Y, Mitsui K, Motizuki M, Tsurugi K. The mechanism of action of ricin and related toxic lectins on eukaryotic ribosomes. The site and the characteristics of the modification in 28 S ribosomal RNA caused by the toxins. *Journal of Biological Chemistry* 1987; 262(12): 5908–5912.

[10] Youle RJ, Murray GJ, Neville DM. Studies on the galactose binding site of ricin and the hybrid toxin man6p-ricin. *Cell* 1981; 23(2): 551-559.

[11] Stirpe F, Batelli MG. Review: Ribosome-inactivating proteins: progress and problems. *Cellular and Molecular Life Sciences* 2006; 63(16): 1850–1866.

[12] Olsnes S, Fernandez-Puemtes C, Carrasco L, Vazquez D. Ribosome inactivation by the toxic lectins abrin and ricin. Kinetics of the enzymatic activity of the toxin A-chains. *European Journal of Biochemistry* 1975; 60(1): 281-288.

[13] Atlas RM. Bioterrorism: from threat to reality. *Annual Review of Microbiology* 2002; 56: 167-185.

[14] Deus-de-Oliveira N, Machado OLT. Allergens and toxins of oleaginous plants: problems and solutions. In: Bernardes MAS (ed). *Environmental Impact of Biofuels.* Rijeka: InTech; 2011. p41-66.

[15] Koja N, Shibata T, Mochida K. Enzyme-linked immunoassay of ricin. *Toxicon* 1980; 18(5-6): 611-618.

[16] Godal A, Olsnes S, Pihl AJ, Radioimmunoassays of ricin and abrin in blood. *Journal of Toxicology and Environmental Health* 1981; 8(3): 409-417.

[17] Ramakrishnan S, Eagle MR, Houston LL. Radioimmunoassay of ricin A- and B-chains applied to samples of ricin A-chain prepared by chromatofocusing and by DEAE Bio-Gel A chromatography. *Biochimica et Biophysica Acta* 1982; 719(2): 341-348.

[18] Poli MA, Rivera VR, Hewetson JF, Merrill GA. Detection of ricin by colorimetric and chemiluminescence ELISA. *Toxicon* 1994; 32(11): 1371-1377.

[19] Shyu HF, Chiao DJ, Liu HW, Tang SS. Monoclonal antibody-based enzyme immunoassay for detection of ricin. *Hybrid Hybridomics* 2002; 21(1): 69-73.

[20] Garber EAE, Eppley RM, Stack ME, McLaughlin MA, Park DL. Feasibility of immunodiagnostic devices for the detection of ricin, amanitin, and T-2 toxin in food. *Journal of Food Protection* 2005; 68(6): 1294-1301.

[21] Ogert RA, Brown JE, Singh BR, Shriver-Lake LC, Ligler FS. Detection of *Clostridium botulinum* toxin A using a fiber optic-based biosensor. *Analytical Biochemistry* 1992; 205(2): 306-312.

[22] Ogert RA, Shriver-Lake LC, Ligler FS. Toxin detection using a fiber optic-based biosensor. In: Lakowics JR, Thompson RB (ed). *Advances in Fluorescence Sensing Technology*; 1993. p11-17.

[23] Narang U, Anderson GP, Ligler FS, Buranst J. Fiber optic-based biosensor for ricin. *Biosensors & Bioelectronics* 1997; 12(9-10): 937-945.

[24] Shyu RH, Shyu HF, Liu HW, Tang SS. Colloidal gold-based immune-chromatographic assay for detection of ricin. *Toxicon* 2002; 40(3): 255-258.

[25] Thullier P, Griffiths G. Broad recognition of ricin toxins prepared from a range of Ricinus cultivars using immune-chromatographic tests. *Clinical Toxicology* 2009; 47(7): 643-650.

[26] Gatto-Menking DL, Yu H, Bruno JG, Goode MT, Miller M, Zulich AW. Sensitive detection of biotoxoids and bacterial spores using an immunomagnetic electrochemiluminescece sensor. *Biosensors & Bioeletronics* 1995; 10(6-7): 501-507.

[27] Yu H, Raymonda JW, McMahon TM, Campagnari AA. Detection of biological threat agents by immunomagnetic microsphere-based solid phase fluorogenic- and electrochemiluminescence. *Biosensors & Bioelectronics* 2000; 14(10-11): 829-840.

[28] Shankar K, Zeng K, Ruan C, Grimes CA. Quantification of ricin concentrations in aqueous media. *Sensors and Actuators B* 2005; 107(2): 640-648.

[29] Wadkins RM, Golden JP, Pritsiolas LM, Ligler FS. Detection of multiple toxic agents using a planar array immunosensor. *Biosensors & Bioelectronics* 1998; 13(3-4): 407-415.

[30] Delehanty JB, Ligler FS. A microarray immunoassay for simultaneous detection of proteins and bacteria. *Analytical Chemistry* 2002; 74(21): 5681-5687.

[31] Weingart OG, Gao H, Crevoisier F, Heitger F, Avondet MA, Sigrist H. A bioanalytical platform for simultaneous detection and quantification of biological toxins. *Sensors* 2012; 12(2): 2324-2339.

[32] Sano T, Smith CL, Cantor CR. Immune-PCR: very sensitive antigen detection by means of specific antibody-DNA conjugates. *Science* 1992; 258(5079): 120-122.

[33] Lubelli C, Chatgiglialoglu A, Bolognesi A, Strocchi P, Colombatti M, Stirpe F. Detection of ricin and other ribosome-inactivating proteins by an immune-polymerase chain reaction assay. *Analytical Biochemistry* 2006; 355(1): 102-109.

[34] He X, McMahon S, Henderson TD, Griffey SM, Cheng LW. Ricin toxic kinetics and its sensitive detection in mouse sera or feces using immune-PCR. *Plos One* 2010; 5(9): e12858. doi: 10.1371/journal.pone.0012858.

[35] Fredriksson SA, Hulst AG, Artursson E, Jong AL, Nilsson C, Baar BLM. Forensic identification of neat ricin and of ricin from crude castor bean extracts by mass spectrometry. *Analytical Chemistry* 2005; 77(6): 1545-1555.

[36] Duriez E, Fenaille F, Tabet JC, Lamourette P, Hilaire D, Becher F, Ezan E. Detection of ricin in complex samples by immunocapture and matrix-assisted laser desorption/ionization time-of-flight mass spectrometry. *Journal of Proteome Research* 2008; 7(9): 4154-4163.

[37] Kull S, Pauly D, Störmann B, Kirchner S, Stämmler M, Dorner MB, Lasch P, Naumann D, Dorner BG. Multiplex detection of microbial and plant toxins by immunoaffinity enrichment and matrix-assisted laser desorption/ionization mass spectrometry. *Analytical Chemistry* 2010; 82(7): 2916-2924.

[38] James W. Aptamers. In: Meyers RA (ed). Encyclopedia of Analytical Chemistry; 2000. p4848-4871.

[39] Kirby R, Cho EJ, Gehrke B, Bayer T, Park YS, Neikirk DP, Mc Devitt JT, Ellington AD. Aptamer-based sensor assay for the detection and quantitation of proteins. *Analytical Chemistry* 2004; 76(14): 4066-4075.

[40] Cho EJ, Collett JR, Szafranska AE, Ellington AD. Optimization of aptamer microarray technology for multiple protein targets. *Analytica Chimica Acta* 2006; 564(1): 82-90.

[41] Haes AJ, Giordano BC, Collins GE. Aptamer-based detection and quantitative analysis of ricin using affinity probe capillary electrophoresis. *Analytical Chemistry* 2006; 78(11): 3758-3764.

[42] Winterbottom DA, Naraynaswamy R, Raimundo Jr IM. Cholesteric liquid crystals for detection of organic vapours. *Sensors and Actuators B* 2003; 90(1-3): 52-57.

[43] Zhao YB, Yu JH, Zhao HF, Tong CY, Wang PH. A novel method for label-free detection of ricin using liquid crystals supported on chemically functionalized surfaces. *Sensors and Actuators B* 2011; 155(1): 351-356.

[44] Anandan S, Anil Kumar GK, Ghosh J, Ramachandra KS. Effect of different physical and chemical treatments on detoxification of ricin in castor cake. *Animal Feed Science and Technology* 2005; 120(1-2): 159-168.

[45] Godoy MG, Gutarra MLE, Maciel FM, Felix SP, Bevilaqua JV, Machado OLT, Freire DMG. Use of a low-cost methodology for biodetoxification of castor bean waste and lipase production. *Enzyme and Microbial Technology* 2009; 44(5): 317-322.

[46] Lugnier AAJ, Le Meur MA, Gerlinger P, Dirheimer G. Inhibition of *in vitro* protein synthesis in a rabbit reticulocyte cell-free system by toxic tryptic peptides from ricin. *Biochimie* 1974; 56(9): 1287-1289.

[47] Barbieri L, Bolognesi A, Cenini P, Falasca AI, Minghetti A, Garofano L, Guicciardi A, Lappi D, Miller SP, Stirpe F. Ribosome-inactivating proteins from plant cells in culture. *Biochemical Journal* 1989; 257(3): 801-807.

[48] Langer M, Rothe M, Eck J, Möckel B, Zinke H. A nonradioactive assay for ribosome-inactivating proteins. *Analytical Biochemistry* 1996; 243(1): 150-153.

[49] Mei Q, Fredrickson CK, Lian W, Jin S, Fan ZH. Ricin detection by biological signal amplification in a well-in-a-well device. *Analytical Chemistry* 2006; 78(22): 7659-7664.

[50] Zamboni M, Brigotti M, Rambelli F, Montanaro L, Sperti S. High-pressure-liquid-chromatography and fluorimetric methods for the determination of adenine released from ribosomes by ricin and gelonin. *Biochemistry Journal* 1989; 259(3): 639-643.

[51] Heisler I, Keller J, Tauber R, Sutherland M, Fuchs H. A colorimetric assay to quantitation of free adenine applied to determine the enzymatic activity of ribosome inactivating proteins. *Analytical Biochemistry* 2002; 302(1): 114-122.

[52] Keener WK, Rivera VR, Young CC, Poli MA. An activity-dependent for ricin and related RNA *N*-glycosidase based on electrochemiluminescence. *Analytical Biochemistry* 2006; 357(2): 200-207.

[53] Cho YC, Keener WK, Garber EAE. Application of deadenylase electrochemiluminescence assay for ricin to food in a plate format. *Journal of Food Protection* 2009; 72(4): 903-906.

[54] Keener WK, Rivera VR, Cho CY, Hale ML, Garber EAE, Poli MA. Identification of the RNA N-glycosidase activity of ricin in castor bean extracts by an electrochemiluminescence-based assay. *Analytical Biochemistry* 2008; 378(1): 87-89.

[55] Hines HB, Brueggemann EE, Hale ML. High-performance liquid chromatography-mass selective detection assay for adenine released from a synthetic RNA substrate by ricin A chain. *Analytical Biochemistry* 2004; 330(1): 119-122.

[56] Becher F, Duriez E, Volland H, Tabet JC, Ezan E. Detection of functional ricin by immunoaffinity and liquid chromatography-tandem mass spectrometry. *Analytical Chemistry* 2007; 79(2): 559-565.

[57] Kalb SR, Barr JR. Mass spectrometric detection of ricin and its activity in food and clinical samples. *Analytical Chemistry* 2009; 81(6): 2037-2142.

[58] Li XP, Bericevic M, Saidasan H, Tumer NE. Ribosome depurination is not sufficient for ricin-mediated cell death in *Saccharomyces cereviseae*. *Infection and Immunity* 2007; 75(1): 417-428.

[59] Brzezinski JL, Craft DL. Evaluation of an *in vitro* bioassay for the detection of purified ricin and castor bean in beverages and liquid food matrices. *Journal of Food Protection* 2007; 70(10) 2377-2382.

[60] Sehgal P, Khan M, Kumar O, Vijayaraghavan R. Purification, characterization and toxicity profile of ricin isoforms from castor beans. *Food and Chemical Toxicology* 2010; 48(11): 3171-3176.

[61] Sehgal P, Kumar O, Kameswararao M, Ravindran J, Khan M, Sharma S, Vijayaragha-van R, Prasad GBKS. Differential toxicity profile of ricin isoforms correlates with their glycosilation levels. *Toxicology* 2011; 282(1-2): 56-67.

[62] Khalil CN, Leite L. Process for producing biodiesel fuel using triglyceride-rich olea-ginous seed directly in a transesterification reaction in the presence of an alkaline al-koxyde catalyst. *US Patent 7*, 112, 229; 2006.

[63] Godoy MG, Fernandes KV, Gutarra MLE, Melo EJT, Castro AM, Machado OLT, Freire DMG. Use of Vero cell line to verify the biodetoxification efficiency of castor bean waste. *Process Biochemistry* 2012; 47(4): 578-584.

[64] Fernandes KV, Deus-de-Oliveira N, Godoy MG, Guimarães ZAS, Nascimento VV, Melo EJT, Freire DMG, Dansa-Petretski M, Machado OLT. Simultaneous allergen in-activation and detoxification of castor bean cake by treatment with calcium com-pounds and solid-state fermentation. *Brazilian Journal of Medical and Biological Research* 2012; 45 in press.

[65] Stirpe F, Pession-Brizzi , Lorenzoni E, Strocchi P, Motanato L, Sperti S. Studies on the proteins of the seeds of Croton tiglium and Jatropha curcas. *Biochemical Journal* 1976; 156(1): 1-6.

[66] Qin W, Ming-Xing H, Ying X, Xin-Shen Z, Fang C. Expression of a ribosome inacti-vating protein (curcin 2) in Jatropha curcas is induced by stress. *Journal of Biosciences* 2005. 30(3): 351-357.

[67] Lin J, Chen Y, Xu Y, Yan F, Tang L, Chen F. Cloning and expression of curcin, a ribo-some-inactivating protein from the seeds of Jatropha curcas. *Acta Botânica Sinica* 2003; 45(7): 858-863.

[68] Adam SEI, Magzoub M. Toxicity of Jatropha curcas for goats. *Toxicology* 1975; 4(3): 388-389.

[69] Ahmed OMM, Adam SEI. Toxicity of Jatropha curcas in sheep and goats. *Research in Veterinary Science* 1979; 27(1): 89-96.

[70] Abdu-Aguye I, Sannusi A, Alafiya-Tayo RA, Brusnurmath SR. Acute toxicity studies with Jatropha curcas L. *Human Toxicology* 1986; 5(4): 269-274.

[71] El Badwi SMA, Mousa HM, Adam SEI, Hapke HJ. Response of Brown Hissex chicks to low levels of *Jatropha curcas, Ricinus communis* or their mixture. *Veterinary & Human Toxicology* 1992; 34: 304-306.

[72] El Badwi SMA, Adam SEI, Hapke HJ. Comparative toxicity of *Ricinus communis* and *Jatropha curcas* in Brown Hissex chicks. *Deutsch Tierarztl Wochenchr* 1995; 102: 75-77.

[73] Beutler JA, Ada AB, McCloud TG, Cragg GM. Distribution of phorbol ester bioactivi-ty in the euphorbiaceae. *Phytoterapic Response* 1989; 3(5): 188-192.

[74] Goel G, Makkar HPS, Francis G, Becker K. Phorbol esters: Structure, biological activi-ty and toxicity in animals. *International Journal of Toxicology* 2007; 26(4): 279-288.

[75] Clemens MJ, Tryner I, Menaya J. The role of protein kinase C isoenzymes in the regulation of cell proliferation and differentiation. *Journal of Cell Science* 1992; 103(4):
881-887.

[76] Brodie C, Blumberg PM. Regulation of cell apoptosis by protein kinase C. *Apoptosis*
2003; 8(1): 19-27.

[77] Adolf HJ, Opferkuch J, Hecker E. Irritant phorbol derivates from four Jatropha species. *Phytochemistry* 1948; 23(1): 129-132.

[78] Rakshit KD, Darukeshwara J, Rathina Raj K, Narasimhamurthy K, Saibaba P, Bhagya
S. Toxicity studies of detoxified Jatropha meal (*Jatropha curcas*) in rats. *Food and Chemical Toxicology* 2008; 46(12): 3621-3625.

[79] Li CY, Devappa RK, Liu JX, Lv JM, Makkar HPS, Becker K. Toxicity of *Jatropha curcas*
phorbol esters in mice. *Food and Chemical Toxicology* 2010; 48(2): 620-625.

[80] Makkar HPS, Becker K, Sporer F, Wink M. Studies of nutritive potential and toxic
constituents of differents provenances of *Jatropha curcas*. *Journal of Agricultural and
Food Chemistry* 1997; 45(8): 3152-3157.

[81] Makkar HPS, Oderibigbe AO, Becker K. Comparative evaluation of non-toxic and
toxic varieties of *Jatropha curcas* for chemical composition, digestibility, protein degradability and toxic factors. *Food Chemistry* 1998; 62(2): 207-215.

[82] Makkar HPS, Siddhuraju P, Becker K. *A laboratory manual on quantification of plant secondary metabolites*. New Jersey: Humana Press; 2007. p130.

[83] Devappa RK, Rajesh SK, Kumar V, Makkar HPS, Becker K. Activities of *Jatropha curcas* phorbol esters in various bioassays. *Ecotoxicology and Environmental Safety* 2012;
78: 57-62.

[84] Martínez-Herrera J, Siddhuraju P, Francis G, Dávila-Ortíz G, Becker K. Chemical
composition, toxic/antimetabolic constituents, and effects of different treatments on
their levels, in four provenances of *Jatropha curcas* L. from Mexico. *Food Chemistry*
2006; 96(1): 80-89.

[85] Devappa RK, Swamylingappa B. Biochemical and nutritional evaluation of Jatropha
protein isolate prepared by steam injection heating for reduction of toxic and antinutritional factors. *Journal of the Science of Food and Agriculture* 2008; 88(5): 911-919.

[86] Joshi C, Mathur P, Khare SK. Degradation of phorbol esters by Pseudomonas aerugi-
nosa PseA during solid-state fermentation of deoiled *Jatropha curcas* seed cake. *Bioresource Technology* 2011; 102(7): 4815-4819.

[87] Barros CRM, Ferreira LMM, Nunes FM, Bezerra RMF, Dias AA, Guedes CV, Cone
JW, Marques GSM, Rodrigues MAM. The potential of white-rot fungi to degrade
phorbol esters of *Jatropha curcas* L. seed cake. *Engineering in Life Sciences* 2011; 11(1):
107-110.

[88] Phasukarratchai N, Tontayakom V, Tongcumpou C. Reduction of phorbol esters in *Jatropha curcas* L. pressed meal by surfactant solutions extraction. Biomass and Bioenergy 2012; 45: 48-56.

[89] Haas W, Sterk H, Mittelbach M. Novel 12-deoxy-16-hydroxiphorbol diester isolated from the seed oil of *Jatropha curcas*. *Journal of Natural Products* 2002; 65: 1434-1440.

[90] Basha SD, Francis G, Makkar HPS, Becker K, Sujatha M. A comparative study of biochemical traits and molecular markers for assessment of genetic relationships between *Jatropha curcas* L. germoplasm from different countries. *Plant Science* 2009; 176(6): 812-823.

[91] He W, King AJ, Khan MA, Cuevas JA, Ramiaramanana D, Graham IA. Analysis of seed phorbol-ester and curcin content together with genetic diversity in multiple provenances of *Jatropha curcas* L. from Madagascar and Mexico. *Plant Physiology and Biochemistry* 2011; 49(10): 1183-1190.

[92] Makkar HPS, Maes J, Greyt WD. Removal and degradation of phorbol esters during pre-treatment and transesterification of *Jatropha curcas* oil. *Journal of the American Oil Chemist's Society* 2009; 86(2): 173-181.

[93] Devappa RK, Makkar HPS, Becker K. Optimization of conditions for the extraction of phorbol esters from Jatropha oil. *Biomass and Bioenergy* 2010; 34(8): 1125-1133.

[94] Ichihashi K, Yuki D, Kurokawa H, Igarashi A, Yajima T, Fujiwara M, Maeno K, Sekiguchi S, Iwata M, Nishino H. Dynamic analysis of phorbol esters in the manufacturing process of fatty acid methyl esters from *Jatropha curcas* seed oil. *Journal of the American Oil Chemist's Society* 2011; 88(6): 851-861.

[95] Katole S, Saha SK, Sastry VRB, Lade MH, Prakash B. Intake, blood metabolites and hormonal profile in sheep fed processed Jatropha (*Jatropha curcas*) meal. *Animal Feed Science and Technology* 2011; 170(1-2): 21-26.

[96] Wang H, Chen Y, Zhao Y, Liu H, Liu J, Makkar HPS, Becker K. Effects of replacing soybean meal by detoxified *Jatropha curcas* kernel meal in the diet of growing pigs on their growths, serum biochemical parameter and visceral organs. *Animal Feed Science and Technology* 2011; 170(1): 141-146.

[97] Becker K, Makkar HPS. Effects of phorbol esters in carp (Cyprinus carpio L.). *Veterinary & Human Toxicology* 1998; 40(2): 82-86.

[98] Kumar V, Akinleye AO, Makkar HPS, Angulo-Escalante MA, Becker K. Growth performance and metabolic efficiency in the Nile tilapia (*Oreochromis* niloticus L.) fed on a diet containing *Jatropha platyphylla* kernel meal as a protein source. *Journal of Animal Physiology and Animal Nutrition* 2012; DOI: 10.1111/j.1439-0396.2010.01118.x.

[99] Yasuraoka K, Hashiguchi J, Blas BL. Laboratory assessment of the molluscicidal activity of the plant *Jatropha curcas* against Oncomelania snail. *Proceedings of the Philippine-Japan joint conference onschistosomiasis research and control*. Manila: 1980. p110-112.

[100] El Kheir YM, El Tohami MS. Investigation of moluscicidall activity of certain Sudanese plants used in folk-medicine. *Journal of Tropical Medicine and Hygiene* 1979; 82(11-12): 237-241.

[101] Liu SY, Sporer F, Wink M, Jourdane J, Henning R, Li YL, Ruppel A. Anthraquinones in Rheum palmatum and Rumex dentatus (Polygonaceae), and phorbol esters in *Jatropha curcas* (Euphorbiaceae) with molluscicidal activity against the schistosome vector snails Oncomelania, Biomphalaria and Bulinus. *Tropical Medicine and International Health* 1997; 2(2): 179-188.

[102] Devappa RK, Makkar HPS, Becker K. Biodegradation of *Jatropha curcas* phorbol esters in soil. *Journal of the Science of Food and Agriculture* 2010; 90(12): 2090-2097.

[103] Rug M, Ruppel A. Toxic activities of the plant *Jatropha curcas* against intermediate snail hosts and larvae of schistosomes. *Tropical Medicine and International Health* 2000; 5(6): 423-430.

[104] Verma M, Pradhan S, Sharma S, Naik SN, Prasad R. Efficacy of karanjin and phorbol ester against termites (Odontotermes obesus). *International Biodeterioration & Biodegradation* 2011; 65(6): 877-882.

[105] Devappa RK, Angulo-Escalante MA, Makkar HPS, Becker K. Potential of using phorbol esters as an insecticide against Spodoptera frugiperda. *Industrial Crops and Products* 2012; 38: 50-53.

[106] Ruebhart DR, Cock IE, Shaw GR. Brine shrimp bioassay: importance of correct taxonomic identification of Artemia (Anostraca) species. *Environmental Toxicology* 2008; 23(4): 555-560.

[107] Kinghom AD, Harjes KK, Doorenbos NJ. Screening procedure for phorbol ester using brine shrimp (Artemia salina) larvae. *Journal of Pharmaceutical Sciences* 1977; 66(9): 1362-1363.

[108] Chumkaew P, Karalai C, Ponglimanon C, Chantraprommat C. Antimycobacterial Activity of phorbol esters from the fruits of *Sapium indicum*. *Journal of Natural Products* 2003; 66(4): 540-543.

[109] Saetae D, Santornsuk W. Antifungal activities of ethanolic extract from *Jatropha curcas* seed cake. *Journal of Microbiology and Biotechnology* 2009; 20(2): 319-324.

[110] Wink M, Grimm C, Koschmiedes C, Sporer F, Bergeot O. Sequestration of phorbol esters by the aposematically coloured bug *Pachycoris klugii* (Heteroptera: Scutelleridae) feeding on *Jatropha curcas* (Euphorbiaceae). *Chemoecology* 2000 ; 10(4): 179-184.

Chapter 4

Bio-Detoxification of Jatropha Seed Cake and Its Use in Animal Feed

Maria Catarina Megumi Kasuya,
José Maria Rodrigues da Luz,
Lisa Presley da Silva Pereira, Juliana Soares da Silva,
Hilário Cuquetto Montavani and
Marcelo Teixeira Rodrigues

Additional information is available at the end of the chapter

1. Introduction

Biodiesel production using the seed oil of *Jatropha curcas* L. (physic nut), as a raw material, results in large amounts of solid residue, called Jatropha seed cake. This seed cake contains lignocellulosic compounds, water, minerals and proteins [1-3]. However, it also contains toxic compounds and anti-nutritional factors [1-3]. The detoxification and reuse of this seed cake is very important for adding economic value, and also reduces potential environmental damage caused by improper disposal of this by-product.

The toxicity of Jatropha seed is mainly attributed to a group of diterpene esters called phor-bolesters. These esters are present in high concentrations in toxic seed varieties but in lower concentrations in a non-toxic seed variety from Mexico [4]. Phorbol esters activate protein kinase C, a key signal transduction enzyme released in response to various hormones and developmental processes in most cells and tissues [5,6].

In addition to the phorbol esters, there is also a toxic protein called curcin in Jatropha seed cake. This protein has two polypeptide chains and is able to inhibit protein synthesis [7]. Curcin is a ribosome-inactivating protein and promotes mucosal irritation and gastrointestinal hemagglutinating action [8].

Phytic acid (myo-inositol hexaphosphoric acid) and tannins are considered anti-nutritional factors because they inhibit the absorption of proteins and minerals [9-11]. Phytic acid is a compound formed during seed maturation [12]. The seed of *J. curcas* has a high concentration of phytic acid, up to 10% of its dry matter [2]. Tannins are polyphenols water-soluble and polar solvents [13]. The tannin content in the seeds of *J. curcas* is low, representing only 3% of its dry weight [13].

Detoxification of the Jatropha seed cake could allow its use as a protein-rich dietary supplement in the animal feed [1,14,15].

The use of residue or by-products in animal nutrition can minimize expenditures on the development of food sources, such as soybean, cotton and wheat meals, without causing undesirable effects on the overall production system. However, it is first necessary to know the nutritional value and effects of the by-product's inclusion in animal diets.

Some studies have used physical and chemical treatments to detoxify Jatropha seed [2,16,17]. These methods have been effective but require the use of chemicals that may result in other the presence of other residues. Conversely, bio-detoxification does not require the application of any chemical compounds. It may also reduce the concentrations of phorbol esters and anti-nutritional factors to non-toxic levels [18].

2. Methodology

2.1. Microorganism, fungal growth conditions and inoculum production (spawn)

The isolate Plo 6 of *P. ostreatus* used in this study belongs to a culture collection from the Department of Microbiology at the Federal University of Viçosa, MG, Brazil. *P. ostreatus* was grown in a Petri dish containing potato dextrose agar culture medium at pH 5.8 and incubated at 25 °C. After seven days, the mycelium was used for inoculum production (spawn) in a substrate made of rice grains [19]. The rice was cooked for 30 min in water with a ratio of 1:3 rice: water (w/w). After cooking, the rice was drained and supplemented with 0.35% $CaCO_3$ and 0.01% $CaSO_4$. Seventy grams of rice was packed into small glass jars and sterilized in an autoclave at 121 ºC for 1 h. After cooling, each jar was inoculated with 4 agar discs (each 5 mm in diameter) containing the mycelium. The jars were then incubated in the dark at room temperature for 15 d.

2.2. Substrate and inoculation

The *J. curcas* seed cake was obtained from an industry of biodiesel (Fuserman Biocombustíveis, Barbacena, Minas Gerais State, Brazil).

To select the most suitable substrates for lignocellulolytic enzyme production, we conducted preliminary experiments with Jatropha seed cake and various lignocellulosic residues. We tested *P. ostreatus* growing on Jatropha seed cake with different percentages of eucalyptus sawdust, eucalyptus bark, corncobs, and coffee husks [20]. The addition of these agroindustrial residues was necessary to balance the carbon and nitrogen ratio, which might benefit mycelial growth [21-23].

The compositions selected for biological detoxification were based on the results of the above preliminary experiments (Table 1). The substrates were humidified with water to 75% of their retention capacity. Then, 1.5 kg of each substrate was placed in polypropylene bags and autoclaved at 121 °C for 2 h. After cooling, the substrates were inoculated with 75 g of spawn and incubated at 25 °C. Samples from non-inoculated autoclaved bags were kept as controls.

Substrates	Mass substrates (kg)	
	Jc	Agroindustrial residue
Jatropha seed cake (Jc)	20	0
Jc + 10% eucalypt bark (JcEb10)	18	2

Table 1. Substrate compositions used for *Pleurotus ostreatus* growth

2.3. Chemical composition of the substrates and enzymatic assays

The phorbol ester contents were analyzed by high performance liquid chromatography (HPLC), as previously described [2]. A standard curve was made using solutions of phorbol-12-myristate 13-acetate (Sigma Chemical, St. Louis, USA) at concentrations from 0.005 to 0.5 mg mL^{-1}.

To determinate the dry mass, 1.5 kg of the substrate was dried at 105 °C until a constant weight was obtained.

The levels of tannins and phytic acid were quantified by a colorimetric method [24,25].

The laccase and manganese peroxidase activities were measured using 2,2'-azino-bis-3-etil-benzotiazol-6-sulfonic acid [26] and phenol red solution [27] as substrates, respectively. Xylanase and cellulase activity was calculated by measuring the levels of reducing sugars produced by the enzymatic reactions [28,29]. Phytase activity (myo-inositol hexakisphosphate phosphohydrolase) was determined using the Taussky-Schoor reagent [30].

The level of reducing sugars was determined by the dinitrosalicylic acid (DNS) method (99.5% dinitrosalicylic acid, 0.4% phenol and 0.14% sodium metabisulfite).A standard curve was made with D-glucose, with concentrations from 0.5 to 1.5 g L^{-1} [31].

2.4. Digestibility of Jatropha seed cake and ammonium production in rumen liquid measured *in vitro*

To analyze the suitability of the chosen substrates (Table 1) in animal feed, we determined their levels of dry matter (DM), organic matter (OM), crude protein (CP), mineral matter (MM), ether extract (EE), neutral detergent fiber (NDF), acid detergent fiber (ADF), acid detergent lignin (ADL), non-fiber carbohydrates(NFC), hemicellulose (HEM), cellulose (CEL) and lignin according to previously described methodology [32,33].

The *in vitro* dry matter digestibility (IVDMD) was determined according to a previous method [34], with some modifications. One liter of rumen fluid was collected from fistulated cattle kept at the Department of Animal Science, Federal University of Viçosa, about two hours after feeding. The animals' diet consisted primarily of grass and corn silage. The rumen digesta was filtered through 4 layers of gauze and the liquid fraction was stored in capped plastic flasks and refrigerated. Ruminal fluid was incubated at 39 °C for 30 min to suspend feed particles and precipitate protozoa, which allowed ruminal bacteria to be collected anaerobically from the middle of the flasks. The IVDMD assay was performed in two steps. In the first step, 350 mg of each substrate sample, harvested before and after colonization by *P. ostreatus*, was incubated with a mixture of 4 mL of ruminal fluid and 32 mL of McDougall buffer. This procedure was performed in anaerobic bottles with a continuous flow of carbon dioxide (CO_2). Bottles were then sealed with rubber stoppers and aluminum closures and incubated for 48 h at 39 °C at 120 rpm. In the second step, after incubation, the filtered material was placed inpre-dried and weighed porcelain filters and washed with hot water four times or until complete removal of all McDougall solution. Next, we added 70 mL of a detergent solution,and the samples were autoclaved for 15 minutes at 121 °C. After heat treatment, the filters were washed again with hot water until complete removal of the detergent solution, and then washed using 10 mL of pure acetone. The filters were heated to 105 °C for 16 h or overnight. After that, the filters were placed in a desiccator, and the dried mass was measured on an analytical balance.

To analyze the production of ammonia, the samples were incubated under the same conditions as described for the IVDMD process. These samples were placed into two flasks containing buffer, rumen fluid and the substrates samples (Table 1) harvested before and after fungal colonization. Ammonia quantification was determined using ammonium chloride as an indicator and absorbance was measured in a spectrophotometer (Spectronic 20D) at 630 nm [35].

2.5. Animal assay

The experiment was conducted in theGoat Experimental Section from the Department of Animal Science at the Federal University of Viçosa - MG, BRAZIL. Twenty-four healthy female Alpine goats weighing 20 ±1.5 kg, with a mean age of five months, were used. This experiment was performed after the Jatropha seed cake had been bio-detoxified by *P.ostreatus*Plo6 [20].

2.5.1. Experimental design

The experimental trial lasted 72 days. During the first12 days, animals were allowed to adapt to the experimental diet. The data were collected during the following 60days.

The animals were kept in individual confinement stables (1.5x2.0m) equipped with food and water systems. The stables had fully slatted floors adapted for the total collection of feces and urine. Water was provided *ad libitum*. The daily food consumption was quantified by subtracting the total offered feed.

The diets were formulated to meet the nutritional requirements of goats with a starting ody weight of 20 kg and a daily weight gain of 100g [37]. The feed contained an average of 12% crude protein.

The treatments consisted ofthe detoxified substrates at four levels: 0, 7, 14 and 20% (based on total dry matter) inforagehayTifton-85 (*Cynodon* spp). The ratio of forage: concentrate was 30:70(Table 2).

Ingredient	Bio-detoxified jatropha seed cake (% dry mass)			
	0	7	14	20
Forage hay Tifton-85	33.18	31.26	31.26	31.25
Jatropha seed cake bio-detoxified	0.00	6.97	13.93	19.90
Maize flour	57.20	55.98	51.84	47.11
Soybean meal	8.37	4.57	1.76	0.53
Sodium chloride	0.20	0.20	0.20	0.20
Calcareous	0.95	0.92	0.90	0.90
ADE vitamins	0.08	0.08	0.08	0.08
Micromineral mixture*	0.03	0.03	0.03	0.03
Sodium bicarbonate	0.40	0.40	0.40	0.40
Chemical composition (%)				
Dry mass (DM)	84.65	84.75	84.80	84.81
Crude protein (CP)	12.84	12.05	11.73	11.89
Ether extract (EE)	3.38	3.25	3.06	2.87
Neutral detergent fiber (NDF)	41.00	42.46	44.42	45.94
Lignin	2.55	4.04	5.67	7.06
Calcium	0.19	0.22	0.26	0.30
Phosphorus	0.27	0.28	0.30	0.32
Net energy (NE, Mcal/kg)	1.87	1.72	1.63	1.54

Table 2. The chemical composition and ingredient proportions of the diet

The feed was supplied twice a day as a complete mixture to allow intake of approximately 10% of the offered amount. The amount was based on the intake of the previous day.

To determine *in vivo* digestibility and nitrogen balance, we collected total feces and urine for five days. Ten percent of the total excretion was sampled. Urine was stored in plastic bags containing 20 mL of sulfuric acid (40% v:v).

The fecal metabolic nitrogen level (Nmet.fecal) was calculated according to previously established methods [37]. The amount of undigested nitrogen (Nund) was calculated as the difference between fecal nitrogen (Nfecal) and Nmet.fecal. To determine the fraction of urinary nitrogen of endogenous origin (Nuend), a previously published equation was used [38]. From the difference between the urinary nitrogen and endogenous urinary nitrogen (Nuend) levels, we calculated the exogenous urinary nitrogen (Nuexo). Nitrogen balance (NB) was estimated with the following equation: NB = N ingested - [Nund + Nuexo]. The biological value of protein was calculated according to previous methods [39].

The chemical compositions of feeds, orts and feces as a percentage of DM, MM, OM, CP, EE, NDF, ADL and NFC were determined according to previous methodology [32, 33, 40]. The total protein level in the bio-detoxified Jatropha seed cake was calculated from the total nitrogen content by applying the correction factor 4.38. The net energy (NE) was obtained by a previously reported equation [41].

The blood samples were collected in the morning, before supplying the feed, by jugular puncture and vacuum tubes. Blood was stored with and without the anticoagulant EDTA. After this procedure, the tubes were refrigerated and sent to a laboratory for blood biochemical analysis to determine the hemogram compounds. This analysis included the numbers of erythrocytes, hemoglobin, hematocrit and leukocytes. In the Blood serum was analyzed creatinine, alkaline phosphatase, urea and total protein.

2.6. Statistical analyses

The experiments on phorbol ester degradation, anti-nutritional factors, *in vitro* digestibility of Jatropha seed cake, and production of liquid ammonia in the rumen were of a randomized design with 5 replicates each. The resulting data were subjected to analysis of variance (ANOVA), and the mean values were compared by Tukey's test ($p < 0.05$) using Saeg software (version 9.1, Federal University of Viçosa).

The experiments on animals were distributed in a completely randomized design with six replicates per diet condition. The resulting data were subjected to analysis of variance (ANOVA) and regression analysis ($p < 0.05$). Regression models (linear, quadratic or cubic) were fitted to the observed significance (5% level of probability) using the REG procedure (SAS 9.0).

3. Results

After 15 days of inoculation, *P. ostreatus* Plo 6 completely colonized the substrates (Figure 1). This illustrates the ability of this fungus to grow in the presence of both phorbol esters and anti-nutritional factors.

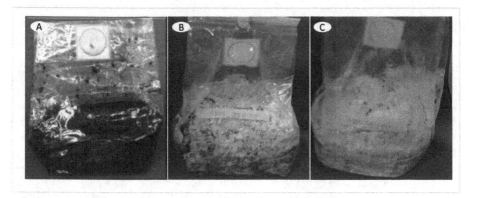

Figure 1. Mycelial growth of *P. ostreatus* Plo 6 in substrate containing varying percentages of Jatropha seed cake. Before the inoculation(A) and 15 days after inoculation: (B) Jatropha seed cake (Jc) and (C)Jc + 10% eucalyptus bark (JcEb10).

3.1. Phorbol ester degradation

Autoclaving the substrates (at 121 °C) reduced the phorbol ester content by an average of 20% (Figure 2). However, these compounds were not degraded at 160 °C for 30 min [17]. Moreover, the addition of sodium hydroxide and sodium hypochlorite combined with heat treatment was able to reduce only 25% of the phorbol concentration [42].

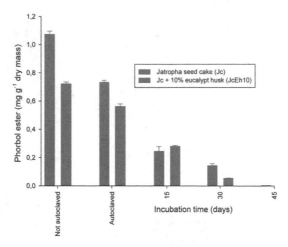

Figure 2. Phorbol ester degradation by *Pleurotus ostreatus* Plo 6 in substrates with varying proportions of Jatropha seed cake (Table 1).

In this study, *P. ostreatus* degraded 99% of the phorbol ester after a 45-day incubation (Figure 2). This rate of degradation was higher than rates observed when chemical deodorization, de-acidification, or bleaching agents were applied to *J.curcas* oil and seed cake [43]. With the exception of bleaching, none of the above chemical processes were effective in reducing the amount of phorbol esters in *J. curcas* seed [44].

The ability of *P. ostreatus* to depolymerize lignin (Figure 3) explains the observed phorbol ester degradation (Figure 2). The degradation of other organic compounds such as chlorophenols and aromatic hydrocarbons also occurs due to depolymerization by laccase and MnP activity [45,46]. The activities of these enzymes of *Phanerochaete* sp [47] and *P. ostreatus* [48] have also been reported to cause dye discoloration in the textile industry and the elimination of pollutants. However, other enzymes may have also influenced degradation of the toxic compounds. Higher cellulase and xylanase activities (Figure 3) were observed between the 15th and 30th incubation days, as indicated by a 58% and 85% degradation of phorbol ester, respectively. However, on the 15th day of incubation we observed lower phorbol ester degradation and lower ligninase activity in the substrate containing eucalyptus bark (Figures 3). This result supports the hypothesis that phorbol ester degradation occurs because of co-metabolism by the enzymes responsible for lignin depolymerization.

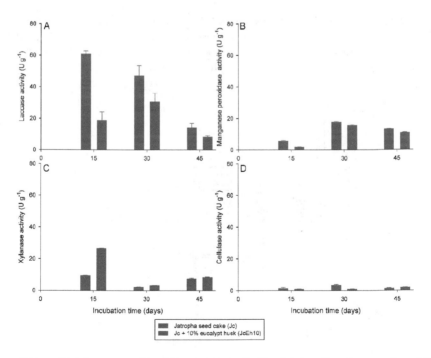

Figure 3. Lignocellulolytic enzymes activity of *Pleurotus ostreatus* Plo 6 in substrates with varying proportions of Jatropha seed cake (Table 1).

After 45 d of substrate incubation with *P. ostreatus*, the residual average phorbol ester concentration was1.8 x 10^{-3} mg g^{-1} dry mass (Figure 2). This concentration is much lower than the 0.09 mg $g^{-1}$of phorbol esters found in the non-toxic variety of *J. curcas* [17].

3.2. Degradation of anti-nutritional factors

Tannin concentrations observed in the seed cake (Figure 4) are similar to those previously reported in the fruit peel of *J. curcas* [4]. The greatest concentration of this compound was observed in the eucalyptus bark substrate (Figure 4). This may have been due to the prior presence of tannins in the eucalyptus bark [49].

The thermal treatment of the substrates decreased the tannin concentration by 46% (Figure 4). This result was similar to that observed in vegetables after cooking or autoclaving at 121 °C and 128 °C for different periods of time [10].

Regardless of the substrate, tannin degradation by *P. ostreatus* Plo 6 increased as a function of the incubation time. The highest observed rate was between 15 and 30 d in the substrate with eucalyptus bark (Figure 4). A high tannin degradation rate was also observed in *Pleurotus* sp. cultivated in coffee husk for 60 d [50].The degradation of tannin is related to tannase activity (tannin acyl hydrolase). This enzyme's activity in polyphenol degradation has been reported in *Aspergillus* and *Penicillium* [51]. Thus, *P. ostreatus* can degrade the tannins in Jatropha seed cake.

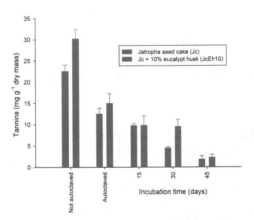

Figure 4. Tannin degradation by *Pleurotus ostreatus* Plo 6 in substrates with different proportions of Jatropha seed cake.

Although phytic acid is considered to be heat-stable [52], the amount of phytic acid decreased by 20% after sterilization of the substrates, (Figure 5). A degradation of 50% of this anti-nutritional factor was also been observed in legumes subjected to autoclaving at 121 ºC for 90 min [10].

Phytase activity by *P. ostreatus* caused a 95% decrease of phytic acid in the substrates (Figure 5). A high degradation rate of this anti-nutritional factor by microbial phytase has previously observed in culture medium containing rapeseed meal that has phytic acid content between 2% to 4 % of the dry mass [53]. The presence of this enzyme has also been observed in *Aspergillus* sp [9], *Agaricus* sp, *Lentinula* sp and *Pleurotus* sp [54]. Thus, *P. ostreatus* degrades the phytic acid that is present in Jatropha seed cake and thereby increases its potential for use in animal feed.

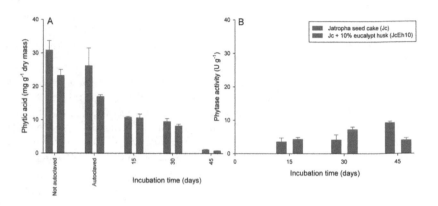

Figure 5. Phytic acid degradation (A) and phytase activity (B) by *Pleurotus ostreatus* Plo 6 in substrates with varying proportions of Jatropha seed cake.

3.3. Digestibility of Jatropha seed cake and ammonium production in rumen liquid *in vitro*

Many agro-industrial residues contain a higher content of fibers, of low digestibility, than proteins, vitamins and minerals. The colonization or fermentation of these by-products by microorganisms, especially lignocellulosic fungi, can efficiently and affordably increase their digestibility and nutritional value [55]. This procedure has been used successfully in cotton waste [56] by colonization with *Brachiaria* sp [57].

Before fungal colonization, we observed higher levels of CP, lignin, ADF and EE in the Jatropha seed cake (Table 3). These data show the importance of adding eucalyptus bark to balance carbon and nitrogen and decrease the fat content, thus resulting in improved fungal growth. Furthermore, these data confirm the potential of using the bio-detoxified seed cake as a source of protein and lipids in ruminant diets [58]. The use of foods rich in these nutrients in animal diets is important because (a) the proteins are the main source of nitrogen and amino acids, and (b) lipids can reduce the production of methane by the rumen [59]. For every 1% increase in the amount of fat added to the diet, there is a 6% reduction in methane emissions by ruminant animals. This reduction in methane production may be due to a negative effect on the lipid protozoa and methanogenic archaea [60].

In the ruminant diet, proteins and amino acids supply nitrogen for microbial protein production. Proteins synthesized by microorganisms of the rumen have a higher nutritional value than dietary protein. According to Alemawor et al. [61], the low level of protein in the skin of cocoa limits its use as animal feed. In this context, increasing the CP in Jatropha seed cake by colonization with *P. ostreatus* (Table 3) increases its potential for use in animal feed.

Components (g 100g⁻¹)	Jatropha seed cake (Jc)		Jc + 10% eucalypt bark (JcEb10)	
	0 (control)	45 days	0 (control)	45 days
Dry mass (DM)	95.027aB	96.243A	95.028aB	96.076A
Organic matter (OM)	93.304aA	91.136B	92.738bA	91.972B
Crude protein (CP)	11.438aB	13.158A	11.075bA	11.264A
Ether extract (EE)	17.929aA	7.563B	16.214bA	7.097B
Non-fiber carbohydrates (NFC)	63.937bB	70.915A	65.259aB	73.800A
Neutral detergent fibre (NDF)	49.217aB	53.920A	49.445aB	54.129A
Acid detergent fibre (ADF)	37.549aA	35.243B	34.442bB	37.363A
Lignin	20.890aA	16.558B	16.902bA	12.246B
Hemicellulose	21.669bA	14.279B	25.331aA	17.022B
Cellulose	25.661bA	23.837B	32.058aA	22.030B
Ash	6.696bB	8.864B	7.262aB	8.028A
In vitro digestibility	54.902bB	77.918A	60.306aB	83.899A

Table 3. The chemical composition of different proportions of Jatropha seed cake and agro-industrial residues colonized for 45 days by *P. ostreatus*

Ether extract content in substrates also decreased after incubation with *P. ostreatus*(Table 3). The reduction was independent of substrate and averaged57%, suggesting that the fungus may have used lipids as a nutrient source. This reduction also contributes to the use of Jatropha residue in ruminant diets because it is typically recommended that EE represent less than10% of a diet's dry matter.

After inoculation with *P. ostreatus* Plo 6, the substrates showed an increase of DM and CP and a reduction of organic material (Table 3). This is similar to observations made in cocoa husks fermented with *P. ostreatus* [61] and in *Jatropha curcas* kernel cake fermented with *Aspergillus niger* and *Tricholoma longibrachiatum* [62]. These data suggest two biological processes: (a) the uptake or absorption of organic matter by the fungus, resulting in the production of proteins and mycelial growth (increased dry weight); and (b) the mineralization or degradation of organic matter resulting in an increase in mineral content (ash) and NFC (Table 3). Assimilation of organic material resulting in an increase in crude protein was also found in

P. sajor-caju grown in cotton waste [56]. Degradation or mineralization of organic matter was also observed in a culture of *P. ostreatus* on eucalyptus bark [63].

We observed an increase in carbohydrates after inoculation with *P. ostreatus* (Table 3). This increase confirms the degradation of more complex compounds such as lignin, cellulose and hemicellulose by *P. ostreatus* (Table 3). Degradation of these compounds in the Jatropha seed cake substrate contributed to an increase in dry matter digestibility (Table 3). An increase in digestibility after lignin degradation has also been shown after cultivation of *Phanerochaete chrysosporium* on cotton stalks [64], *P sajor-caju* in agroindustrial residue [57] and *P. ostreatus* on eucalyptus bark [63].

Therefore, colonization of Jatropha seed cake by *P. ostreatus*, with or without addition of eucalyptus bark, was shown to increase the nutritional value and *in vitro* digestibility of this by-product from the biodiesel production chain.

3.3.1. Ammonia production by microorganismsin the ruminal liquid

The ruminant's microorganisms are large and genetically diverse, consisting of bacteria, fungi, protozoa and viruses [65]. These microorganisms contribute to the fermentation of substrates that have low solubility (e.g., plant material rich in fiber) in organic acids, methane, ammonia, acetate, lactate, formate, ethanol, propionate, CO_2 and H_2 [66].

The ammonia production observed in this study can be considered low (Figure 6). The production of this compound by rumen ammonia-producing bacteria may vary from 33 to 159% of the dry mass depending on the bacterial species and protein content of the diet [67]. Rumen microorganisms are capable of incorporating a large portion of the produced ammonia by deamination of amino acids and hydrolysis of nitrogen compounds. However, when the rate of deamination exceeds the rate of assimilation, protein catabolism results in the undesirable and inefficient process of high ammonia production and low retention of nitrogen [68]. This undesirable process can be characterized by the loss of protein through the excretion of nitrogen as urea in the animal's urine [69]. According to previous studies, the rate of dietary protein degradation is directly proportional to the ammonia production and protein nitrogen loss. Therefore, low ammonia production by microorganisms in the rumen fluid demonstrates that the substrates colonized with *P. ostreatus* exhibit good digestibility and that the rate of ammonia uptake is not exceeded (Figure 6).

The highest ammonia production was observed in substrates colonized by *P. ostreatus* (Figure 6). The increased production of ammonia may be due to the higher amount of crude protein in these substrates (Table 3). This shows that the colonized substrates have a greater capacity to be degraded by rumen microorganisms than those that were not colonized. This higher capacity can be a result of the following: (a) the presence of toxic compounds which inhibit microbial growth in substrates without fungal colonization or (b) a reduction in lignocellulosic compounds by enzymatic action resulting in compounds (e.g., NFC, CP and NDF) contributing to the growth and metabolism of microorganisms (Table 3). This degradation/mineralization of lignocellulosic compounds is also reflected in the increase in dry matter and ash content (Table 3).

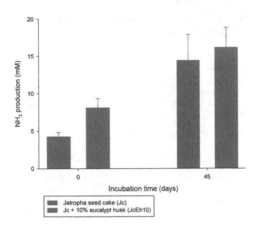

Figure 6. Ammonia concentration in batch cultures containing ruminal liquid and the substrates evaluated (Table 1). Substrates were harvested before and after 45 days of colonization by *Pleurotus ostreatus* Plo 6.

Finally, it is important to note that *P. ostreatus* Plo 6's reduction in levels of phorbol esters (99%) and ammonia (Figure 6) did not inhibit the development of rumen bacteria. This result confirms the detoxification of Jatropha seed cake by *P. ostreatus* and again highlights the importance of fungal colonization in the preparation of the cake for use in animal feed.

3.4. Animal assay

3.4.1. Food intake, digestibility and nitrogen balance

The intake of dry matter and nutrients was influenced by the different amounts of detoxified Jatropha seed cake in the diet (Table 4). The DM intake (% BW) and NDF showed a quadratic response (P < 0.05), and there was a positive linear effect (P < 0.05) on DM intake (g/kg BW0.75). The DM, OM and CP increased linearly (P < 0.05) and no changes were observed in either the EE and NFC consumption by animals (Table 4).

The increase in DM intake may be attributed to a reduction of the energy values in the experimental diets (Table 2). In this sense, the animals ate more DM to reach their energy requirements. Thus, we can infer that the consumption and palatability of the diets was not restricted by the inclusion of detoxified Jatropha seed cake, although it increased DM intake by the animals. In prior experimental animals, the replacement of soybean meal by Jatropha seed cake resulted in a decrease in DM ingestion, which was attributed to the presence of anti-nutritional factors [70]. In this study, the maximum intake of DM and NDF was 3.68 and 1.67% of BW, respectively. This was not enough to promote the rumen fill effect. In diets with a low energy level, animals tend to exceed the consumption limit of 1.2% of BW, offsetting any food energy deficiency [66].

The increase in the DM intake resulted in increases in intakes of OM and CP. However, this increase had no effect on the overall consumption of EE and NFC. These results support the theory of compensation in DM intake by animals on diets with a low concentration of energy.

Intake (g d⁻¹)	Jatropha seed cake (% dry mass)				Regression	R^2	CV (%)
	0	7	14	20			
Dry mass (DM)	701.34	719.90	826.27	891.35	$Y = 681.609+10.058x$	0.94	13.67
DM (%BW)	2.91	2.84	3.19	3.68	$Y = 2.901-0.030x+0.003x^2$	0.99	10.33
DM (g/kgBW$^{0.75}$)	64.41	63.61	71.89	81.65	$Y = 61.327+0.883x$	0.83	10.42
Neutral detergent fiber(%BW)	1.17	1.19	1.38	1.67	$Y = 1.170-0.009x+0.001x^2$	0.99	9.83
Organic matter	673.35	696.55	795.80	854.01	$Y = 657.092+9.545x$	0.95	13.62
Crude protein	91.40	87.55	97.29	106.09	$Y = 87.469+0.791x$	0.72	14.10
Ether extract	23.54	24.10	26.10	26.06	ns	--	14.07
Non-fiber carbohydrates	272.74	274.36	304.92	307.44	ns	--	16.10
Digestibility (%)							
Dry mass	74.16	68.10	65.40	62.46	$Y = 73.328-0.565x$	0.97	7.24
Organic matter	74.88	69.02	65.90	63.06	$Y = 74.136-0.577x$	0.97	6.83
Crude protein	67.92	54.10	52.14	46.70	$Y = 65.305-0.984x$	0.89	10.53
Ether extract	80.40	77.62	78.61	76.92	ns	--	5.18
Neutral detergent fiber	69.32	64.19	60.55	58.05	$Y = 68.774-0.560x$	0.98	7.44
Non-fiber carbohydrates	82.46	78.50	75.75	73.75	$Y = 82.050-0.432x$	0.98	7.99

Table 4. Consumption and apparent total tract digestibility of dry matter and nutrients in goats fed with biodetoxified Jatropha seed cake

The inclusion of increasing levels of detoxified Jatropha seed cake promoted a linear reduction ($P < 0.05$) in the digestibility of DM, OM, CP, NDF and NFC diets tested (Table 4). The exception was EE digestibility, which did not show significant variation and had average values of 78.39%. This reduction in dry matter digestibility can be attributed to an increase in passage rate as a function of consumption, resulting in the shorter digestion time of nutrients in the gastrointestinal tract [66]. This effect is associated with the highest possible lignin concentration of the experimental diets.

In relation to nitrogen metabolism, significant effects on Nuendo, nitrogen balance and the biological value of protein from the level of detoxified Jatropha seed cake added were not observed (Table 5). The intake of nitrogen, excretion of Nfecal, Nmet.fecal, Nundig, Nuexo and urinary nitrogen were influenced in a linear manner at the levels studied (Table 5). Losses of nitrogen in the urine and feces were 28.63 and 40.20% of the consumed nitrogen, respectively.

Variable (g d^{-1})	Jatropha seed cake (% dry mass)				Regression	R^2	CV (%)
	0	7	14	20			
Consumed nitrogen	14.62	14.01	15.57	16.98	Y = 13.993+0.126x	0.72	14.10
Nfecal	4.67	5.22	6.79	7.95	Y = 4.416+0.169x	0.96	20.44
Nmet. fecal	0.39	0.37	0.42	0.45	Y = 0.372+0.003x	0.72	14.14
Nund	4.28	4.84	6.37	7.50	Y = 4.042+1.166x	0.97	21.64
Urinary nitrogen	4.56	3.49	4.29	5.19	Y= 4.495-0.194x+0.011x^2	0.70	25.38
NUend	1.80	1.86	1.90	1.80	Ns	--	8.07
Nuexo	2.77	1.63	2.39	3.39	Y = 2.703-0.213x+0.012x^2	0.75	43.50
NB	7.58	7.53	6.80	6.09	ns	--	36.41
BVP (%)	72.54	77.91	72.70	63.86	ns	--	21.18

Table 5. Nitrogen consumption, excretion, balance and retention by goats fed with bio-detoxified Jatropha seed cake

Generally, urea concentration is correlated with ammonia content in ruminants because digestive microorganisms using nitrogen require energy for the synthesis of bacterial proteins. Most likely, there was excess of ruminal ammonia, which increased the excretion of nitrogen in the urine; thus, levels of 12% CP in the diet of growing goats can promote higher levels in waste nitrogen. Valadares et al. [71] also found an increase in nitrogen excretion in urine when they provided a similar amount of protein to zebu cattle.

Nitrogen balance (NB) and biological value did not differ between the evaluated diets. However, the positive observed values of NB suggest its use in the synthesis of tissue.

3.4.2. Blood parameters

The experimental diets did not significantly alter the blood parameters of the animals (Table 6). The resulting values were similar to those of normal goats [72]. The hemoglobin concentration was similar to that observed in goats fed with Jatropha seed cake [62].

From the leukocyte values observed in this study, it could be inferred that animals did not experience inflammation after ingesting bio-detoxified Jatropha seed cake (Table 6).

The absence of significant effects in the content of creatinine, alkaline phosphatase and total protein by the different levels of bio-detoxified Jatropha seed cake (Table6) shows that liver function was not altered in animals fed the experimental diets.

Variable	Jatropha seed cake (% dry mass)				Regression	CV (%)	Reference*
	0	7	14	20			
Hematological profile							
Erythrocytes (x 10^6 mm^3)	4.93	5.16	4.81	5.01	ns	11.88	8 a 18
Hemoglobin (g dL^{-1})	11.61	11.55	11.60	11.82	ns	5.93	8 a 12
Hematocrit (%)	31.75	31.83	31.68	32.33	ns	6.18	22 a 38
Leukocytes (n µL^{-1})	12179	12405	10533	11583	ns	14	4000 a 13000
Biochemical profile							
Creatinine (mg dL^{-1})	0.76	0.82	0.76	0.75	ns	7.48	1 a 1.82
Alkaline phosphatase (U l^{-1})	173.95	154.90	186.62	163.84	ns	18.95	93 a 387
Urea (mg dL^{-1})	20.76	21.53	18.09	20.53	ns	14.47	21.4 a 42.8
Total proteins (g dL^{-1})	6.90	6.66	6.73	6.57	ns	4.71	6.4 a 7

Table 6. Hematological and biochemical blood profiles of goats fed with bio-detoxified Jatropha and of normal control goats

Urea levels in the blood can increase in response to diets with low energy [73]. However, we did not observe this effect.

Thus, inclusion of Jatropha seed cake bio-detoxified by *P. ostreatus* shows promise as an animal feed supplement because it did not result in changes to blood parameters or clinical symptoms of poisoning. This included goats fed with up to 20% of the treated residue. Conversely, ingestion of Jatropha residue treated with organic solvents (ethanol and hexane) caused diarrhea and other side effects in swine [74]. Similarly, diarrhea and death resulted in goats fed with Jatropha seed cake colonized by *Aspergillus*, *Penicillium* or *Trichoderma* [62].

4. Conclusions

The residue of *J. curcas* increases with increased biodiesel production, so it is necessary to find an appropriate usefor these residues. In this study we demonstrate the potential to transform the residue of biodiesel containing lignocelluloses, toxic compounds, and anti-nutritional factors into animal feed. This process adds economic value to biodiesel production and avoids theimproper disposal of its by-products in the environment.

Thebio-detoxification of Jatropha seed cake promotes the reduction of phorbol ester levels and increases the nutritional value of this residue. The resulting alternative food can be included in amounts up to20% (DM)in the diet of growing goats.

Acknowledgments

The authors thank Coordenação de Aperfeiçoamento de Pessoal de Nível Superior (CAPES), Conselho Nacional de Desenvolvimento Científico e Tecnológico (CNPq), and Fundação de Amparo à Pesquisa do Estado de Minas Gerais (FAPEMIG) for financial support; Biovale Energy for support; and Fuserman Biocombustíveis for kindly donating the Jatropha seed cake. Additionally, the authors thank all friends and collaborators of the Laboratory of Mycorrhizal Association and the Department of Microbiology of the Federal University of Viçosa.

Author details

Maria Catarina Megumi Kasuya[1*], José Maria Rodrigues da Luz[1],
Lisa Presley da Silva Pereira[2], Juliana Soares da Silva[1], Hilário Cuquetto Montavani[2] and Marcelo Teixeira Rodrigues[2]

1 Department of Microbiology, Federal University of Viçosa, Viçosa, Minas Gerais, Brazil

2 Department of Animal Science, Federal University of Viçosa, Viçosa, Minas Gerais, Brazil

References

[1] Gubitz GM, Mittelbech M, Trabi M. Exploitation of tropical oil seed plant *Jatropha curcas* L. Bioresource Technology 1999; 67, 73-82.

[2] Makkar HPS, Becker K, Sporer F, Wink M. Studies on nutritive potential and toxic constituents of different provenances of *Jatropha curcas*. Journal of Agricultural and Food Chemistry 1997;45, 3152-3157.

[3] Openshaw K. A review of *Jatropha curcas*: an oil plant of unfullled promise. Biomass and Bioenergy 2000;19, 1-15.

[4] Makkar HPS, Aderibigbe A0, Becker K. Comparative evaluation of non-toxic and toxic varieties of *Jatropha curcas* for chemical composition, digestibility, protein degradability and toxic factors. Food Chemistry 1998;62, 207-215.

[5]   Slater SJ, Ho C, Stubbs CD. The use of fluorescent phorbol esters in studies of protein kinase C–membrane interactions. Chemistry and Physics of Lipids 2002; 116, 75–91.

[6]   Saraiva L, Fresco P, Pinto E, Gonçalves J. Characterization of phorbol esters activity on individual mammalian protein kinase C isoforms, using the yeast phenotypic assay. European Journal of Pharmacology 2004;491, 101– 110.

[7]   Kumar A, Sharma S. An evaluation of multipurpose oil seed crop for industrial uses (Jatropha curcas L.): A review. Industrial Crops and Products 2008;28, 1-10.

[8]   Barbieri L, Battellim, Stirpe F. Ribosome-inactivating protein from plants. Acta Biochimica Biophysica 1993;1154, 237-282.

[9]   Ullah AHJ, Phillippy BQ. Substrate selectivity in Aspergillus ficuum phytase and acid phosphatases using myo-inositol phosphates. Journal of Agricultural and Food Chemistry 1994;42, 423–425.

[10] Rehman Z, Shah WH. Thermal heat processing effects on antinutrients, protein and starch digestibility of food legumes. Food Chemistry 2005; 91, 327–331.

[11] Liang J, Han BZ, Nout, MJR, Hamer RJ. Effect of soaking and phytase treatment on phytic acid, calcium, iron and zinc in rice fractions. Food Chemistry 2009;115, 789–794.

[12] Torre M, Rodriguez AR, Saura-Calixto F. Effects of dietary fiber and phytic acid on mineral availability. Critical Reviews in Food Science and Nutrition 1991;1,1-22.

[13] Makkar HPS, Becker K. Do tannins in leaves of trees and shrubs from Africa and Himalayan regions differ in level and activity. Agroforestry Systems 1998; 40, 59-68.

[14] Staubmann R, Ncube I, Gübitz GM, Steiner W, Read JS. Esterase and lipase activity in Jatropha curcas L. seeds. Journal of Biotechnology 1999;75, 117-126.

[15] Haas W, Sterk H, Mittelbach M. Novel 12-deoxy-16-hydroxyphorbol diesters isolated from the seed Oil of Jatropha curcas. Journal of Natural Products 2002;65, 1434-1440.

[16] Haas W, Mittelbach M. Detoxification experiments with the seed oil from Jatropha curcas L. Industrial Crops and Products 2000;12, 111-118.

[17] Aregheore EM, Becker K, Makkar HPS. Detoxification of a toxic variety of Jatropha curcas using heat and chemical treatments, and preliminary nutritional evaluation with rats. South Pacific Journal of Natural Science 2003; 21, 50-56.

[18] Universidade Federal de Viçosa. Destoxificação biológica da torta de pinhão manso e a utilização da mesma como ração. Instituto Nacional da propriedade Industrial, BR 10 2010 002912-0, 2012.

[19] de Assunção LS, da Luz JMR, da Silva MCS, Vieira PAF, Bazzolli DMS, Vanetti MCD, Kasuya MCM. Enrichment of mushrooms: an interesting strategy for the acquisition of lithium. Food Chemistry 2012;134, 1123-1127.

[20] Da Luz JMR. Degradação de compostos tóxicos e de fatores antinutricionais da torta de pinhão manso por Pleurotus ostreatus, portuguese (Degradation of toxic compounds

and of antinutritional factors of the Jatropha seed cake by *Pleurotus ostreatus*). Ms, thesis, Universidade Federal de Viçosa; 2009.

[21] Nunes MD, da Luz JMR, Paes SA, Ribeiro JJO, da Silva MCS, Kasuya MCM. Nitrogen supplementation on the productivity and the chemical composition of oyster mushroom. Journal of Food Research 2012;1, 113-119.

[22] Elisashvili V, Penninckx M, Kachlishvili E, Tsiklauri N, Metreveli E, Kharziani T, Kvesitadze G. *Lentinus edodes* and *Pleurotus* species lignocellulolytic enzymes activity in submerged and solid-state fermentation of lignocellulosic wastes of different composition. Bioresource Technology 2012;99, 457–462.

[23] Giardina P, Palmieri G, Fontanella B, Rivieccio V, Sannia G. Manganese peroxidase isoenzymes produced by *Pleurotus ostreatus* grown on wood sawdust. Archives of Biochemistry and Biophysics 2000;376, 171–179.

[24] Makkar HPS, Bluemmel M, Becker K. Formation of complexes between polyvinyl pyrrolidone and polethylene glycol with tannins and their implications in gas production and true digestibility in *in vitro* techniques. British Journal Nutrition, 1995;73, 897–913.

[25] Gao Y, Shanga C, Maroof MAS, Biyashev RM, Grabau EA, Kwanyuen P, Burton JW, Buss GR. A modofied colorimetric method for phytic acid analysis in soybean. Crop Science 2007;47, 1797-1803.

[26] Buswell J, Cai YJ, Chang ST. Effect of nutrient nitrogen and manganese or manganese peroxidase and laccase production by *Lentinula (Lentinus) edodes*. Fems Microbiology Letter 1995;128, 81-88.

[27] Kuwahara M, Glenn JK, Morgan MA. Separation and characterization of two extracellular H_2O_2-dependent oxidases from ligninolytic cultures of *Phanerochaete chrysosporium*. FEMS Microbiology Letters 1984;169, 247-250.

[28] Mandels M, Andreotti R, Roche C. Measurement of saccharifying cellulase. Biotechnology and Bioengineering 1976;6, 21-33.

[29] Bailey MJ, Biely P, Pantonen K. Interlaboratory testing of methods for assay of xylanase activity. Journal of Biotechnology, 1992;23, 257-270.

[30] Harland BF, Harland J. Fermetative reduction of phytate in rye, white, and whole wheat breads. Cereal Chemistry 1980;57, 226-229.

[31] Cavallazzi JRP, Brito MS, Oliveira MGA, Villas-Bôas SG, Kasuya MCM. Lignocellulolytic enzymes profile of three *Lentinula edodes* (Berk.) Pegler strains during cultivation on eucalyptus 317 bark-based médium. Food, Agriculture and Environment 2004;2, 291-297.

[32] Silva DJ, Queiroz AC. Análise de alimentos (métodos químicos e biológicos). Viçosa: Universidade Federal de Viçosa; 2002.

[33] Van Soest PJ, Robertson JB, Lewis BA. Methods of dietary fiber, neutral detergent fiber, and nonstarch polysaccharides in relation to animal nutrition. Journal of Animal Science, 1991;74, 3583-3597.

[34] Tilley JM, Terry RA. A two- stage technique for the *in vitro* digestion of forage crops. Jounal of the British Grassland Society, 1963;18,104-111.

[35] Chaney AL, Marbach EP. Modified reagents for determination of urea and ammonia. Clinical Chemistry 1962;8, 130-132.

[36] National Research Council – NRC. Nutrient requirements of small ruminants. Washington: National Academy of Sciences; 2007.

[37] Moore JE, Goetsch AL, Luo J, Owens FN, Galyean ML, Johnson ZB, SahluT, Ferrell CL Prediction of fecal crude protein excretion of goats. Small Ruminant Nutrition 2004;53, 275-292.

[38] Coelho da Silva JF, Leão MI. Fundamentos de nutrição de ruminantes. Piracicaba: Livro; 1979.

[39] Rossi Jr P, Pereira JRA. Manual prático de avaliação de alimentos. Piracicaba: FEALQ; 1995.

[40] Moe PW, Flatt WP, Tyrrel HF. Net energy value of feeds for lactation. Journal Dairy Science 1972;55, 945.

[41] Goel G, Makkar HPS, Francis G, Becker K. Phorbol esters: structure, biological activity, and toxicity in animals. International Journal of Toxicology 2007;26, 279-288.

[42] Rakshit KD, Darukeshwara J, RAJ KR, Narasimhamurthy K, Saibaba P, Bhagya S. Toxicity studies of detoxified Jatropha meal (*Jatropha curcas*) in rats. Food and Chemical Toxicology 2008;46, 3621–3625.

[43] Ahmed WA, Salimon J. Phorbol ester as toxic constituents of tropical *Jatropha curcas* seed oil. European Journal of Scientific Research 2009;31, 429-436.

[44] Barr DP, Aust SD. Mechanisms white-rot fungi use to degrade pollutants. Environmental Science and Technology, v. 28, p. 78-87, 1994.

[45] Majcherczyk A, Johannes C, Hüttermann A. Oxidation of polycyclic aromatic hydrocarbons (PAH) by laccase of Trametes versicolor. Enzyme and Microbial Technology 1998;22, 335-341.

[46] Perez, J., De La Rubia T., Ben Hamman, O., Martinez, J. *Phanerochaete flavido-alba* laccase induction and modification of manganese peroxidase isoenzyme pattern in decolorized olive oil mill wastewaters. Applied and Environmental Microbiology 1998; 64, 2726-2729.

[47] Iamarino G, Rao MA, Gianfreda L. Dephenolization and detoxification of olive-mill wastewater (OMW) by purified biotic and abiotic oxidative catalysts. Chemosphere 2009; 74, 216–223.

[48] Vázquez G, González-Alvarez J, Santos J, Freire MS, Antorrena G. Evaluation of potential applications for chestnut (*Castanea sativa*) shell and eucalyptus (*Eucalyptus globulus*) bark extracts. Industrial Crops and Products 2009;29, 364–370.

[49] Fan L, Soccol AT, Pandey A, Vandenberghe LPS, Soccol CR. Effect of caffeine and tannins on cultivation and fructification of *Pleurotus* on coffee husks. Brazilian Journal of Microbiology, 2006;37, 420-424.

[50] Batra A, Saxena RK. Potential tannase producers from the genera *Aspergillus* and *Penicillium*. Process Biochemistry 2005;40, 1553–1557.

[51] Deshpande SS, Damodaran S. Food legumes: chemistry and technology. Advances in Cereal Science and Technology 1990;10, 147-241.

[52] El-Batal AI, Karem A. Phytase production and phytic acid reduction in rapeseed meal by *Aspergillus niger* during solid state fermentation. Food Research International 2001;34, 715–720.

[53] Collopy PD, Royse DJ. Characterization of phytase activity from cultivated edible mushrooms and their production substrates. Journal of Agricultural and Food Chemistry, 2004;52, 7518-7524.

[54] Villas-Bôas S, Esposito E, Mitchell DA. Microbial conversion of lignocellulosic residues for production of animal feeds. Animal Feed Science and Technology 2002;98, 1-12.

[55] Castro ALA, Paiva PCA, Dias ES, Santos J. Avaliação das alterações bromatológicas e de degradabilidade do resíduo de lixadeira do algodão após tratamento biológico com *Pleurotus sajor-caju*. Ciências Agrotécnicas 2004;28, 608-613.

[56] Bisaria R, Madan M, Vasudevan P. Utilization of agro residues as animal feed through bioconversion. Bioresource Technology 1996;59, 5-8.

[57] Martínez-Herrera J, Siddhuraju P, Francis G, Dávila-Ortíz G, Becker K. Chemical composition, toxic/antimetabolic constituents, and effects of different treatments on their levels, in four provenances of *Jatropha curcas* L. from México. Food Chemistry 2006;96, 80–89.

[58] Mohammed N, Onodera R, Itabashi H, Lila ZA. Effects of ionophores, vitamin B6 and distiller's grains on in vitro tryptophan biosynthesis from indolepyruvic acid, and production of other related compounds by ruminal bacteria and protozoa. Animal Feed Science and Technology 2004;116,301-311.

[59] Dohme F, Machmuller A, Kreuzer, M. Ruminal methanogenesis as Influenced by Individual fatty acids supplemented to complete ruminant diets. Letters in Applied Microbiology 2001;32, 47-51.

[60] Alemawor F, Dzogbefia VP, Oddoye EOK, Oldham JH. Effect of *Pleurotus ostreatus* fermentation on cocoa pod husk composition: Influence of fermentation period and Mn^{2+} supplementation on the fermentation process. African Journal of Biotechnology, 2009;8, 1950-1958.

[61] Belewu MA, Akande BA. Biological upgrading of the nutritional quality of *Jatropha curcas* kernel cake: effect on performance characteristics of goat. International Research Journal of Biotechnology 2010;1, 19-22.

[62] Bento CP. Composição química e digestibilidade *in vitro* de resíduos agroindustriais tratados com fungos da podridão branca. Ms, thesis. Universidade Federal de viçosa; 2008.

[63] Shi J, Chinn MS, Sharma-Shivappa RR. Microbial pretreatment of cotton stalks by solid state cultivation of *Phanerochaete chrysosporium*. Bioresource Technology 2008;99, 6556-6564.

[64] Dehority BA, Orpin CG. Development of, and natural fluctuations in, rumen microbial populations. In: Hobson PN, Stewart CS (eds). The Rumen Microbial Ecosystem. London: Chapman and Hall; 1996.

[65] Van Soest PJ. Nutritional ecology of the ruminant. Ithaca: Cornell University; 1994.

[66] Lima CP, Backes C, Villas Bôas RL, de Oliveira MR, Manabe KTA, Freitag, EE . Bermuda grass sod production as related to nitrogen rates. Revista Brasileira de Ciência do Solo 2010;34,371-378.

[67] Wallace RJ, Broderick GA, Brammall ML. Microbial protein and peptide metabolism in rumen fluid from faunated and ciliate-free sheep. British Journal Nutrition 1987; 58:87-93.

[68] Russell JB, O'connor JD, Fox DJ. A net carbohydrate and protein system for evaluating cattle diets: I. Ruminal fermentation. Journal of Animal Science 1992;70, 3551-3561.

[69] Amaral PM, Oliveira DM, Campos JMS. Torta de Pinhão manso em substituição ao farelo de soja em dietas para bovinos. Belém: Sociedade Brasileira de Zootecnia; 2011.

[70] Valadares RFD, Gonçalves LC, Sampaio IB. Níveis de proteína em dietas de bovino. 4. Concentrações de amônia ruminal e uréia plasmática e excreções de uréia e creatinina. Revista Brasileira de Zootecnia 1997;26, 270-1278.

[71] Pugh DG. Clínica de ovinos e caprinos. São Paulo: Roca; 2005.

[72] Contreras PA, Wittwer F, Böhmwald H. Uso dos perfis metabólicos no monitoramento nutricional dos ovinos. Porto Alegre: Universidade Federal do Rio Grande do Sul; 2000.

[73] Chivandi E, Erlwanger KH, Makuza SM, Read JS, Mtimuni JP. Effects of dietary *Jatropha curcas* meal on percent packed cell volume, serum glucose, cholesterol and triglyceride concentration and alpha-amylase activity of weaned fattening pigs. Journal of Animal and Veterinary Sciences 2006;1, 18-24.

Biodiesel Applications in Engines

Chapter 5

Application of Biodiesel in Automotive Diesel Engines

Yanfei Li, Guohong Tian and Hongming Xu

Additional information is available at the end of the chapter

1. Introduction

1.1. background

Diesel engines due to the better fuel economy have been widely used in automotive area. However, the limited reserve of fossil fuel and deteriorating environment have made scientists seek to alternative fuels for diesel while keeping the high efficiency of diesel engine. Fuel consumption is expected to increase from 86 million barrels per day to 112 million barrels per day by 2035 according to the report published by US Energy Information Administration in 2011 [1]. The limited reserve cannot afford this usage. Another challenge is environmental deterioration and climate change. Excessive emissions of carbon dioxide (CO_2) to the atmosphere are regarded as the leading cause of global warming. In addition, other emissions, such as NO_x, SO_2, also have a close relationship with other forms of climate change, such as photochemical smog and acid rain. Due to these, the regulations on fuel economy and emission limits are increasingly stringent. Table 1 shows the EU emissions regulations for passenger cars came into force since 1992.

Tier	Date				HC+NO$_x$	
Euro 1†	Jul-92	2.72 (3.16)	-	-	-	0.97 (1.13)	0.14 (0.18)
Euro 2	Jan-96	1	-	-	-	0.7	0.08
Euro 3	Jan-00	0.64	-	-	0.5	0.56	0.05
Euro 4	Jan-05	0.5	-	-	0.25	0.3	0.025
Euro 5	Sep-09	0.5	-	-	0.18	0.23	0.005
Euro 6 (future)	Sep 14	0.5	-	-	0.08	0.17	0.005

Table 1. European emissions regulations for passenger cars (Category M*), g/km

However, it is not a long-term solution even though these measures can help alleviate or re-duce the emissions and extend the lifetime of fossil fuel in industry, because one day fossil fuel would run out if the fuel consumption is kept at nowadays' rate. In addition, the de-crease in fossil fuel reserve would lead to the increase of oil price. The rising fuel price raised the cost-competiveness of other energy sources, such as wind energy and solar ener-gy. Hence some sectors, such as industrial and buildings, are driven towards other substi-tute energy sources when possible, whereas in transportation sector, liquid fuel is still the preferred choice. Consequently, the transportation share of the total liquid fuels increases in the projected period, accounting for 80% of the total increase in liquid fuel production [1].

Therefore, the efforts have been made to seek the alternative for fossil fuel, especially after the energy crisis in 1970s. People are trying to find a sustainable way to power the engines.

1.2. Biodiesel

Among the alternatives for fossil diesel, biodiesel has been widely investigated due to its renewability, comparable properties to fossil diesel and the reduction in main emission products.

Biodiesel is mainly comprised of mono-alkyl esters of long chain fatty acids and it was de-fined in standard ASTM D6751. Normally feedstock such as vegetable oil and animal fat is used to produce biodiesel through transesterification method.

With the on-going development of biodiesel, the categorization of biodiesel is developed. Generally biodiesel can be categorised by the readiness of feedstock and produce technolo-gies. The biodiesel made from vegetable oil and animal fats using transesterification method is normally recognised as first-generation biodiesel. The second-generation biodiesel, Bio-mass to liquid (BTL) fuel is to turn cellulose into fuel components (enzyme fermentation or gasification through Fischer-Tropsch synthesis), and the feedstock theoretically can be any bio mass such as waste agriculture, wood chips etc. Some biodiesel from jatropha, algae, etc., despite being produced by transesterification method, is widely regarded as second-generation due to the technical challenge of feedstock planting and harvesting. Normally the second-generation biodiesel can supplement the drawbacks of the first-generation bio-diesel particularly being non-competitive with food.

Different from above definition, another new fuel, Hydro-treated vegetable oil (HVO), using the same feedstock as 1st generation biodiesel, is viewed as second generation biodiesel, and BTL is third-generation [17]. The authors still categorize it into second-generation biodiesel because HVO shares the same feedstock with first-generation biodiesel even though it is made through different way and has better quality than first-generation biodiesel through transesterification.

1.2.1. History

Vegetable oil has been used in diesel engine long time ago. In 1900 after the invention of die-sel engine,), Dr. Diesel used peanut oil to run one of his engines at the Paris Exposition of 1900. Vegetable oils were used in diesel engines until 1920s. The recent use of vegetable oil as the alternative for diesel starts from early 1980s due to the concern about the energy supply. But biodiesel is not commercialised until late 1990s.For the direct use of vegetable oil, several difficulties occur, including the high viscosity, acid composition, free fatty acid content, and gum formation due to oxidation and polymerization during storage and combustion, carbon deposition, and oil thickening (Ma and Hana, 1999). Therefore, the direct use of vegetable oil may not be satisfactory and practical. The technologies to improve the vegetable oil appeared.

1.2.2. Production process

There are several ways used to produce FAME through vegetable oil, pyrolysis, cracking and transesterification. The common method is transesterification. Figure 1 shows the chemical reaction of FAME production. Triglycerides, the main component in vegetable oil and animal fat, reacts with alcohol in a caustic environment and produce Fatty Acid Methyl Ester (FAME) or Fatty Acid Ethyl Ester (FAEE) and the byproduct glycerol. As a result, biodiesel is a mixture of esters, small amount of glycerol, free fatty acids, partially reacted acylglycerol, and residual raw materials. Normally methanol is used for the reaction for the higher reaction rate and lower price. The fuel qualities may be varied in terms of alcohol used. Methyl ester was better than ethyl ester from the point of engine performance: higher power and torque could be achieved.

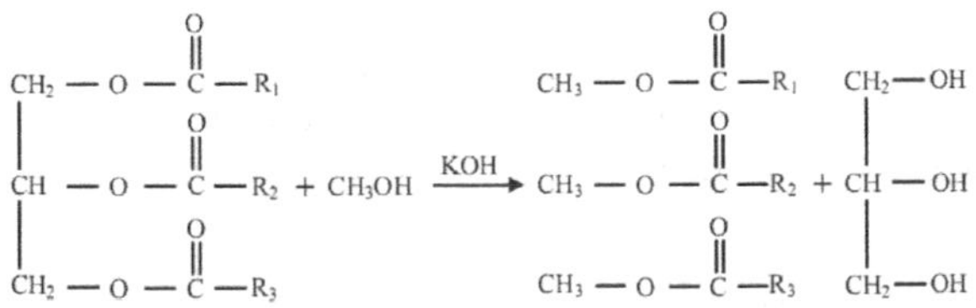

Figure 1. Transesterification reaction in caustic environment

Different from transesterification, Hydro-treating of vegetable oil or animal fat (HVO) have been developed by several companies, such as Neste Oil, Axens IFP, and Honeywell UOP. In the hydro-treating process, vegetable oil or animal fat is also the feedstock. Hydrogen is added into the plant to remove the oxygen content and saturate the C=C and the final products are paraffin, propane, water and CO_2. Propane is also a promising and valuable fuel product. Due to the excellent properties,

$$\text{Triglycerides} + H_2 \xrightarrow[\text{Catalyst}]{\text{Green}} \text{diesel} + H_2O/CO_2 + \text{Propane}$$

Figure 2. Product routes of transesterification and hydro-treated method

Fischer-Tropsch (FT) method is another way to produce synthetic fuel, using various ligno-cellulosic feedstock. BTL (biomass-mass-to-liquid) fuel, GTL (Gas-to-liquid) fuel, and CTL (Coal-to-liquid) fuel are produced with this method. GTL and BTL are not sustainable fuels for natural gas and coal are not renewable. However, in this chapter GTL is included later because it shares the similar production process and has similar physiochemical properties with BTL. Figure 3 shows the manufacturing process of FT synthetic fuel. The solid feedstock (coal and biomass) are initially gasified, then the composition of the syngas and CO_2 and sulphur compounds are removed before the synthesis process. After the synthesis process, the products are refined and the refined products includes the synthetic diesel and gasoline blendstock.

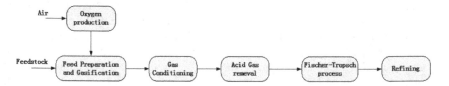

Figure 3. FT fuel manufacturing process

1.3. Biodiesel standards

Due to the difference in the feedstock and manufacturing process, the FAME products may vary very much. Table 2 lists several main standards used in the world, aiming at reach the satisfaction and the equipment compatibility.

Bio-Diesel	Unit	Austrian Standard C1190 Feb. 91	Australian Bio-diesel Standard	DIN 51606 (1997/9/1)	U.S. Quality Specification NBB/ASTM	Euro Standard EN 14214
Density at 15°C	g/cm3	0.86 - 0.90	0.86 – 0.89	0.875 - 0.90	/	0.86 - 0.90
Viscosity at 40°C	mm2/s	6.5 - 9.0 (20°C)	3.5 – 5.0	3.5 - 5.0	1.9 - 6.0	3.50 - 5.00
Flash point	°C	Min. 55	120.0°C	Min. 110	Min. 100	Min. 120
	(°F)	-131		-230	-212	-248
CFPP	°C (°F) summer	Max. 0 (32)	/	Max. 0 (32)	/	
	winter	Max. -8 (17.6)		Max. -20 (-4)		
Total sulphur	mg/kg	Max. 200	Max. 50 mg/kg	Max. 100	Max. 500	Max. 10.0
Conradson (CCR) at 100% at 10%	% mass	Max. 0.1	Max 0.05	Max. 0.05	Max. 0.05	/
		/	Max 0.30	/	/	Max. 0.30

Cetane number	-	Min. 48	Min. 51	Min. 49	Min. 40	Min. 51
Sulfated ash content	% mass	Max. 0.02	Max. 0.02	Max. 0.03	Max. 0.02	Max. 0.02
Water content	mg/kg	free of deposited water		Max. 300	/	Max. 500
Water & sediment	vol. %	/	Max. 0.05	/	Max. 0.05	/
Total contamination	mg/kg	/	Max. 24	Max. 20	/	Max. 24
Copper corrosion (3 hs, 50°C)	degree of Corrosion	/	< 10mg/kg sulphur – 1 "/ 10mg/kg sulphur – 3 max	1	No. 3b max.	1
Neutralisation value	mg	Max. 1	/	Max. 0.5	Max. 0.8	Max. 0.50
Oxidation stability	h	/	Min 6 @ 110°C	/	/	Min. 6.0
Methanol content	% mass	Max. 0.30	0.2	Max. 0.3	Max. 0.2	Max. 0.20
Ester content	% mass	/	Min 96.5	/	/	Min 96.5
Monoglycerides	% mass	/	/	Max. 0.8	/	Max. 0.80
Diglycerides	% mass	/	/	Max. 0.4	/	Max. 0.20
Triglycerides	% mass	/	/	Max. 0.4	/	Max. 0.20
Free glycerine	% mass	Max. 0.03	Max. 0.02	Max. 0.02	Max. 0.02	Max. 0.02
Total glycerine	% mass	Max. 0.25	Max. 0.25	Max. 0.25	Max. 0.24	Max. 0.25
Iodine value		/	/	Max. 115	/	Max. 120
Linolenic acid ME	% mass	/	/	/	/	Max. 12.0
Polyunsaturated ("/ =4db)	% mass	/	/	/	/	Max. 1
Phosphorus content	mg/kg	/	Max. 10	Max. 10	/	Max. 10.0
Alkaline content (Na +K)	mg/kg	/	/	Max. 5	/	Max. 5.0
Alkaline earth metals (Ca + Mg)	mg/kg	/	/	/	/	Max. 5.0

Table 2. Biodiesel Standards

1.4. Pros and Cons

The following summarised the advantages of biodiesel:

- Renewable energy source in comparison with traditional fossil fuel

- Degradability

- Much less sulphur, leading to lower toxic substances in the exhaust

- Absence of PAHs and around 10% of oxygen help the reduction of HC and CO

- Various feedstock

The use of bio-diesel fuels cannot occur without adopting a series of precautions. Indeed, unless the proper precautions are taken, biodiesel fuels can cause a variety of engine performance problems including filter plugging, injector coking, piston ring sticking and breaking, seal swelling and hardening/cracking and severe lubricant degradation. Bio-diesel also requires special treatment at low temperatures to avoid an excessive rise in viscosity and loss of fluidity.

Long-term storage problems can be observed as result of the poor oxidation stability of biodiesel fuels. Thus additives may be needed to improve storage conditions. Furthermore, biodiesel is an excellent medium for microbial growth. As water accelerates microbial growth and is more prevalent in biodiesel than in petroleum based fuels, special care must be taken to remove water from fuel storage tanks to avoid operational problems such as sediment build-up, premature filter plugging or storage tank corrosion.

1.5. Security of supply

Another reason for the search of alternative fuel is the energy security. The economical growth can promote the demand for energy. Table 3 listed crude oil reliance on imported oil of US and China. The reliance of the two countries are up to 44.8% and 56.54%, respectively. The energy supply can be well alleviated if biodiesel can be produced and used in commercial scale. Further analyses are needed to understand the fuel difference and can help fuel design during the biodiesel production process. The Commission Green Paper (CEC, 2000) reported an ambitious EU programme on the usage of biodiesel that 20% alternative fuel substitution by 2020 in conventional fuel in the road transport sector is set. On another hand, the utilisation of biodiesel leads to concerns of land use, deforestation and negative effect on bio-diversity needs further exploration.

	Year 2007	Year 2008	Year 2009	Year 2010	Year 2011
US[1]	58.2%	57.0%	51.5%	49.2%	44.8%
China	47.2%	49.8%	52%	54.8%	56.5%

Table 3. [1] US Department of Energy, Energy Information Administration, Monthly Energy Review, Washington, DC, March 2012, Table 3.3aCrude oil reliance of US and China from 2007 - 2011

2. Fuel properties

2.1. Fuel composition

Due to the various feedstocks for biodiesel, the fuel composition varies in a wide range. Generally the fats and oils contain 10 common types of fatty acid consisting of 12- to 22-carbon chain, and over 90% are between 16- and 18-carbon chains [11]. Table 4 shows the composition of some common FAME. Some of these are saturated, some are monounsaturated and others are poly-unsaturated. The composition of biodiesel determined the chemical and physical properties, such as the fuel viscosity, surface tension, cetane number (CN), oxidation stability, low-temperature properties, as well as the following combustion and emission characteristics.

2.2. Viscosity

Viscosity is a measure of resistance to flow of a liquid due to internal friction and it is one of the most important parameters in evaluate the fuel quality. Viscosity affects engine working process very much. Higher viscosity would prohibit atomisation and instability of fuel droplets, and promote the formation of deposit. This also explains why neat vegetable oils have difficulty when used in diesel engines directly. The viscosity can be measured according to the standards such as ASTM D445 or ISO 3104. The viscosity of individual saturated fatty acid ester increases with carbon chain length and non-linearly decreases with the increase of number of double bonds [4]. In addition, the position of C=C double bond and the branching in the ester moiety has less effect on viscosity. Biodiesel has a higher viscosity than fossil diesel. At lower blend ratio, the viscosities of diesel and biodiesel/diesel blend are very close. As the blend ratio continues to increase, biodiesels show a much higher value. This can partly explain why biodiesel/diesel blends with lower blend ratio are widely used in diesel engines.

		Rapeseed (high erucic)	Rapeseed (low erucic)	Soybean	Sunflower	Coconut oil	Palm kernel oil	Palm oil (Africa)	Palm oil (Indonesia)	Beef tallow	Chicken fat	Fish oil
Saturated fatty acids	< C10:0					~13	~7					trace
	C12:0	trace	trace		trace	41-46	41-45	trace	trace-0.5	trace		trace
	C14:0	trace	trace	<0.5	trace-0.1	18-21	15-17	1-2	~1	2-4	~1	6-9
	C16:0	2-4	3-5	8-12	5.5-8	9-12	7-10	41-46	41-47	23-29	20-24	11-20
	C18:0	1-2	1-2.5	3-5	2.5-6.5	2-4	2-3	4-6.5	4-6	20-35	4-7	1-4
	C20:0	0.5-1	<1	<0.5	<0.5	trace	trace-0.3	~0.5	~0.5	<0.5		trace-1
	C22:0	0.5-2.0	trace-0.5	trace	0.5-1.0	trace	trace-0.5			trace		trace
	C24:0	0.5	trace-0.2		<0.5							trace-1
Unsaturated fatty acids	C14:1					trace				~0.5		trace
	C16:1	~0.5	0.1-0.5	trace	<0.5	trace		<0.5	~0.5	2-4	~7	6-11
	C18:1	11-24	52-66	18-25	14-34	5-9	10-18	37-42	37-41	26-45	38-44	12-15
	C18:2	10-22	17-25	49-57	55-73	0.5-3	1-3	8-12	~10	2-6	18-23	1-2
	C18:3	7-13	8-11	6-11	trace-0.4	trace	trace-0.5	trace-0.5	trace-0.5	~1	~1	0.5-1
	C20:1	~10	1.5-3.5	<0.5	<0.5	trace	trace-0.5	trace	trace	<0.5		1-16
	C20:x*		trace-0.1							trace		6-19
	C22:1	41-52	trace-2.5		trace-0.3							1-20
	C22:x*									~0.5	0.5-1	5-14
	C24:1		trace									trace-1
Other constituents	Phosphatides	2.5		1.1-3.2	<1.5				0.05-0.1		<0.07	
	Tocopherols	0.07-0.08		0.09-0.12	0.07-0.1	0.003			0.02-0.12		0.001	
	Sterols	0.5-1.1		0.2-0.4	0.25-0.45	0.05-0.1	0.08-0.12		0.04-0.08		0.08-0.14	~0.3

* x > 1
trace ≤ 0.05%

http://www.dieselnet.com/tech/fuel_biodiesel_app.html

Table 4. Composition of common FAME.

2.3. Cetane number (CN)

CN is used to evaluate fuel ignition quality determined by the time between start of injection and start of combustion. Higher CN indicates shorter time after the injection. CN is mainly determined by the fuel composition and can affect engine startability, noise and emission characteristics. Generally, biodiesel has a higher CN than mineral diesel. This can be attributed to the longer carbon chain length of biodiesel. Unsaturation and carbon chain length are the most two influential factors of CN [16, 12, 22]. Higher saturation degree and longer fatty acid chain length can lead to a lower CN. The positions of chemical group may also influence the CN. The CNis the highest when the carbonyl group is at the end of the carbon chain and lowest in the middle of the carbon chain. In addition, a higher level of hydroperoxides increases CN and a shorter chain length of the alcohol moiety may also increase CN [12, 22].

2.4. Low-temperature property

Diesel engines may encounter the start-up and performance problems at low temperatures. As ambient temperatures decrease towards the fuel saturation temperature, high-molecular-weight compound begin to nucleate and form was crystals. The existence of was crystals may affect the fuel supply and engine performance. Three parameters, cloud point (CP), pour point(PP), and cold filter plugging point (CFPP) are used to describe low-temperature properties. The temperature at which crystals become visible is called CP because the crystals lead to a cloudy suspension. if the temperature continues to decrease, the crystals would fuse together and form larger agglomerates. The temperature at which crystal agglomeration is large enough to prevent free pouring of fluid is called PP. A more complicated test procedure is involved in order to obtain CFPP. The test uses a vacuum to draw a 20 cc fuel sample through a 45 micron screen within a 60 seconds. The lowest temperature that the fuel can still flow through the filter is called CFPP.

This is an very concerning issue in application of biodiesel into diesel engines. Neat biodiesel has poorer low-temperature performance than conventional diesel. Therefore, when biodiesel is used in cold condition, the biodiesel crystafdl may block the fuel pipe and the fuel filter, and even abrade the high-pressure fuel pump, shorting the lifetime of vehicle engines. Research has shown that the cold flow property is associated with the saturated FAME in vegetable oil based biodiesel. The higher the proportion of saturated FAME and the longer chain FAME in saturated FAME, the poorer the cold temperature performance is [31].

Generally, the low-temperature properties can be improved by following methods:

• Blending with fossil diesel

• blending with additives

• Crystallization fractionation by decreasing the saturated alkyl ester content in the biodiesel.

• Employing branched esters

3. Application in Diesel Engines

The engine performance fuelled with biodiesel is crucial for the application of biodiesel. The mainly involved problems may include corrosion, material degradation, injector coking, filter plugging and piston ring sticking, engine deposits etc. therefore, in the following section, the studies focusing on these issues were introduced.

3.1. Fuel spray characteristics

Injection spray is the process that fuel is injected from nozzle, and it is associated with following fuel atomisation, interaction with surrounding gas, mixture formation and combustion. Regarding to a new fuel applied into the diesel engine, the spray process is different due to the different properties from diesel, and the control strategy should be changed accordingly in order to achieve the optimum performance. Viscosity, surface tension and density are the three main parameters, which influence fuel spray characteristics. Higher viscosity and surface tension will prohibit the atomisation and instability of fuel droplets. Due to the different biodiesels properties from diesel, studies on the spray characteristics are necessary.

3.1.1. Near-field spray characteristics

In the near-field of nozzle, the spray is dominated by the injection dynamics while the spray is affected by the ambient conditions in the far field. According to Hiroyasu's model, before the $t_{breakup}$, which represents the time for fuel jet breakup, the penetration length is proportional to the time after start of injection, namely ASOI. However, the non-linear phenomenon has been observed by a number of researchers. The acceleration process has been found to be different among fuels. Figure 4 compares the morphology of the spray process of the three tested fuels, ULSD, RME and GTL and Figure 5 shows the spray tip penetration length evolution after start of injection (ASOI) using an ultrahigh-speed CCD camera of up to 1 million shots per second. The initial non-linear penetration can be observed, indicating the acceleration period at the initial spray stage. GTL fuel has longer penetrating length than RME and diesel even though it was overtaken by RME 70 μs ASOI. Several publications have reported that GTL with lower density has a shorter penetration delay. However, these were based on the global fuel spray characteristics using a relatively low speed camera [21, 13]. The temporal resolution is not high enough to capture the near-field spray process.

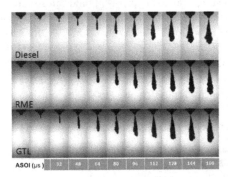

Figure 4. Sequence of spray images in a single time-resolved ULSD spray (Pinj=120 MPa, P_{amb} =3.0 MPa and t_{dur}=1.5 ms)

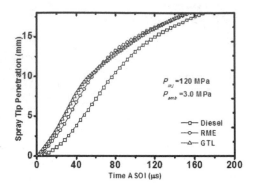

Figure 5. Spray tip penetration length evolution against time ASOI (P_{inj}=120 MPa, P_{amb} =3.0 MPa and t_{dur}=1.5 ms)

3.1.2. Macroscopic spray characteristics

Normally, biodiesel shows a longer penetration and narrower spray angle than fossil fuel due to the higher viscosity, surface tension and density. The penetration length of biodiesel increases with the blend ratio, higher biodiesel content requires longer breakup time [15]. The difference between the two type fuels can be varied at different conditions. Senatore et al. [24] experimentally studied biodiesel spray characteristics at different ambient pressures. The authors showed that little difference can be observed at the ambient pressure of 1.2 MPa while the penetration length significantly increased in contrast to diesel spray at the ambient pressure of 5.0 MPa. In addition, biodiesel may have a lower penetration velocity due to the negative effect of fuel density on spray velocity [9].

3.1.3. Sauter Mean diameter (SMD)

SMD is one of the parameters to evaluate fuel atomisation quality and represents the ratio of total droplet volume to surface area. Smaller SMD indicates more small fuel droplets and the larger contact area with surrounding gas. Due to the high viscosity and surface tension, SMD of biodiesel is higher than fossil diesel. Allen et al. [3] conducted the comparative analysis on 15 biodiesels and a larger SMD, between 5%-40%, can be observed and concluded an empirical equation to estimate SMD:

$$SMD=0.002103\mu + 0.000330\sigma \tag{1}$$

where μ is fuel dynamic viscosity (Pa.s) and σ is fuel surface tension (N/m).

Figure 6 compared diesel with neat RME and GTL at different injection pressure along the spray axis in terms of SMD. It can be seen that the injection pressure has a significant impact on droplet size. The SMD decreases dramatically when the injection pressure increases from 80 MPa to 120 MPa. GTL has the lowest SMD among all the three measured fuels at the given conditions while RME has the largest droplet size. The SMD evolution also decreases with the increase of the axial distance downstream of the nozzle even though there is a slightly increase from 40 mm to 50 mm at the 80 MPa condition. This may be caused by the droplet coalesce.

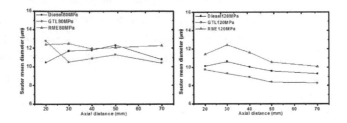

Figure 6. SMD distribution along the spray axis under injection pressure of 80 MPa (Left) and 120 MPa (Right)

3.1.4. Wear Performance and Durance

In diesel engines, the engine parts are lubricated by the fuel itself. In order to meet diesel engine emission standards, Ultra-low sulphur diesel (ULSD) are produced, which has a maximum sulphur content of 15 ppm. However, the relatively poor lubricity of ULSD may lead to the failure of engine parts, such as fuel pumps and injectors. The inherently greater lubricity of Biodiesel can offset the drawback of ULSD, and a small percentage of biodiesel can restore the lubricity of diesel [28].

It is also necessary to study the engine endurance in order to fully apply biodiesel into vehicle operation. Graboski et al. [14] reviewed previous studies and concluded that nitrile rubber, Nylon 6/6 and high-density polypropylene exposed to methyl soyester and D-2 blends exhibited changes in physical properties and fluorinated elastomer must be adopted for biodiesel application. Terry et al. [30] examined the durability of a set of five commonly used elastomers in automotive fuel systems in different biodiesel blends (B5 and B20) and the effect of a highly oxidized biodiesel blends on the elastomers was studied. The results demonstrated that it appeared to be compatible with these elastomers, for highly oxidized and unoxidized B5 and unoxidized B20, but B20 prepared from highly oxidized biodiesel shows the potential for significant problems.

3.2. Engine Output performance

The adaptability of biodiesel in diesel engines has been well studied from low blend ratio to neat biodiesel. Due to the potential damage of biodiesel on vehicle, normally biodiesel blended with diesel were mostly studied. In general, typical heating value for biodiesel is lower than that of fossil diesel. A greater amount of fuel is subsequently required to maintain the same engine output. Greater fuel consumption of up to 13% with heavy-duty engines over the United States Federal Test Procedure (US-FTP) cycle was observed. Due to the lower heating value, engine power loss is expected and the loss increases with the blend ratio of biodiesel in diesel [25, 33]. Figure 7 shows the output power of an 4-cylinder common-rail diesel engine with different biodiesel blends at two engine speeds. With the increase of biodiesel blend concentration, maximum out power was gradually reduced, especially in the higher blend ratio. Figure 8 presents the brake specific fuel consumption (BSFC) corresponding to the condition of Figure 7. The obvious increase in fuel consumption has been observed using higher biodiesel/diesel blends. From Figure 7 and Figure 8, it can be found that the output performance and fuel economy of biodiesel/diesel blends are very close to those of diesel when the blend ratio is under 20%. Therefore, biodiesel/diesel blends with lower blend ratio are preferred.

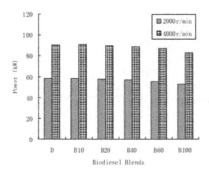

	Power		
	2000r/min	4000r/min	Average
D	0	0	0
B10	-0.1%	0.8%	0.35%
B20	-1.1%	-0.6%	-0.85%
B40	-2.6%	-1.4%	-2%
B60	-4.5%	-2.9%	-3.7%
B100	-9.4%	-7.8%	-8.65%

Figure 7. Output power of different biodiesel blends at two speeds [33]

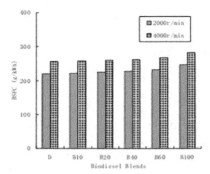

	BSFC		
	2000r/min	4000r/min	Average
D	0	0	0
B10	0.6%	0.4%	0.5%
B20	1.8%	1.2%	1.5%
B40	3.0%	1.8%	2.4%
B60	5.4%	4.3%	4.85%
B100	12.1%	10.0%	11%

Figure 8. BSFC of different biodiesel blends at two speeds [33]

3.3. Combustion characteristics

In diesel engines, combustion is to release energy contained in fuel, then impart work on piston, and power the engine. Factors affecting combustion characteristics include fuel properties and in-cylinder conditions. Biodiesel has a higher CN and the effect of CN on combustion has been discussed in previous section. The average peak cylinder pressure increases when biodiesel or its blends are used. For the application of biodiesel into diesel engines, advanced injection timing and increased injection pressure have been normally used. This is due to their differences in density and bulk modulus of compressibility. Combustion and emissions characteristics have been investigated by Chuepeng et al. [7] using different RME blends from B0 to B50 in a single-cylinder diesel engine in terms of engine load, EGR (Exhaust gas recirculation), and injection timing. At the same engine load, the proportion of fuel burnt in the premixed phase increases and the start of combustion is advanced as the proportion of RME in ULSD increases. With the same operating conditions, increase in EGR rate of up to 20%, slightly reduces the peak pressure and increases ignition delay.

3.4. Emissions

3.4.1. Regulated emissions

A number of studies on the engine emissions of engines powered by biodiesel or blends have been carried out. Environmental Protection Agency (EPA) in the United States correlated the biodiesel ratio with the changes in pollutants using statistical regression analysis and also the average effect of biodiesel on heavy-duty diesel engines [10]. The NO_x emissions increased with the concentration of biodiesel and the increase is by 10% at B100 while HC, CO and PM were greatly reduced. The significant reduction of emissions of HC, CO and PM can be attributed to the oxygen content in biodiesel.

It has been widely reported that NO_x increases as biodiesel is used in diesel engines. A number of efforts have been made in order to understand the formation mechanism and eliminate this penalty. There are several main reasons have been suggested:

- Advanced injection timing

- Oxygen content in biodiesel

- Double bond

- Radiative heat transfer

- Higher adiabatic flame temperature

An advanced injection timing due to the higher bulk modulus of biodiesel in pump-in-line injection system leads to the earlier start of combustion, resulting in higher in-cylinder temperature, which can increase NO_x emission [29, 2]. However, it is well understood that advanced injection timing increases NO_x emission in diesel engine [20], and this seems not to be the main contribution to NO_x increase as broad application of common rail injection system, which can well control the injection timing. Schmidt et al. [23] experimentally studied the effect of concentration of oxygen in intake gas on NO_x emission, and found that NO_x emission increases with the oxygen content in mixture. However, the effect of oxygen content in air on combustion is different from that of oxygen content in biodiesel itself. The radiative heat transfer may also play a role in the NO_x increase. Soot radiation is the primary way of heat loss from in-cylinder flame and biodiesel can reduce this heat loss and will increase the flame temperature and produce more NO_x [6]. The double bond in biodiesel composition is another potential to increase NO_x emission. The double bonds lead to higher adiabatic flame temperature, and the biodiesel with higher unsaturated ester percentage corresponded to higher NO_x emission [27, 19]. Ban-Weiss et al. [5] also revealed that slight difference in the adiabatic flame temperature can lead to a measurable increase in NO_x. Mueller et al. [20] suggested that NO_x increase in biodiesel-fuelled engine is the result of a number of mechanisms, and the relative importance of each mechanism may vary under different operating conditions and indicated that air/fuel mixture close to stoichiometric at ignition and in the standing premixed auto-ignition zone near flame lift-off length may be the key factors in explaining the NO_x increase, whose effect could cause higher local and average in-cylinder temperature and lower radiative heat losses.

Therefore, three main strategies to alleviate the NO_x emission can be proposed: one is to determine the biodiesel compound that can lower NO_x emission or use a proper base fuel and additives, another is to design the combustion system to prohibit NO_x production by lower the combustion temperature, and the third one is to recalibrate the engine by tuning the injection strategy.

3.4.2. Unregulated emissions

For other unregulated emissions from an engine fuelled with biodiesel, polycyclic aromatic hydrocarbon (PAH) and nitro PAH compounds are substantially reduced, as well as the lower levels of some toxic and reactive HC species [26]. The PM composition (i.e. volatile material and elemental carbon) from the combustion of RME-based biodiesel blend (B30) in a turbo-charged engine with EGR operation was studied using thermo-gravimetric analysis (TGA) [8]. Generally, total PM mass from B30 combustion was lower than that for diesel in all engine operating conditions. Elemental carbon PM mass fractions were slightly lower for the B30. The volatile material portions of the B30 particulates are greater than those of diesel particulates irrespective of engine operating condition. For both fuels used in the test, volatile material was observed to be higher at idle speed and light load when exhaust emissions were at low temperature.

In previous regulations on PM, mass is the only concern. With the increasing concern on exhaust emissions, the PM size and number are to be limited by future emission regulations. [32] studied the particulate matter characteristics of RME10 and GTL10. It was found that the application of RME10 and GTL10 leads to a reduction in both total particle number and non-volatile part number over the test conditions. The obtained images from SEM (Scanned Electronic Microscopy) for the three test fuels are shown in Figure 9. The images show the morphology of PM at two magnifications. The authors found that PM from diesel combustion has more clusters than those from RME10 and GTL10 from Figure 10 (a), (c) and (e), indicating that primary particle size of the tested fuels is around 20 mm Figure 10. (b), (d) and (f).

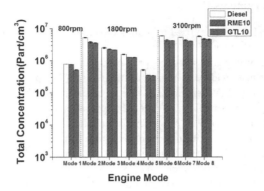

Figure 9. Exhaust particulate number concentration (total)

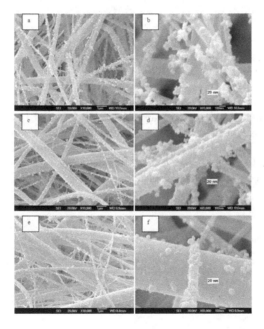

Figure 10. Particle morphology (captured under engine mode of 1800 rpm, 30 Nm): (a) Diesel magnification of 10000; (b) Diesel magnification of 65000; (c) RME 10 magnification of 10000; (d) RME magnification of 65000; (e) GTL10 magnification of 10000; (f) GTL10 magnification of 65000

3.5. Engine emission optimisation

Two popular methods have been used to reduce the engine out emission for biodiesel-fuelled engines: injection strategy and EGR. For the former, the combustion process can be controlled by injection timing and injection pressure. For the time being, the common rail injection system has been widely used and multiple injections up to of 5 times can be realised. Through this way, the fuel injection rate is controllable. The NO_x can be reduced through pre-injection with small amount fuel; this prevents a long period of ignition delay and therefore leads to a lower peak pressure; for the latter, EGR is always an effective way to reduce NO_x emission. Due to the induction of exhaust gas, the global in-cylinder temperature is reduced, avoiding the thermal conditions favoured by NO_x formation. Ladommatos et al. [18] also revealed that the reduction in combustion temperature is a consequence of the reduced peak rate of the premixed phase combustion due to the lower oxygen availability.

4. Conclusions

Biodiesel is the most promising fuel in the near future as an alternative to fossil diesel. Despite of its advantages, it still has some disadvantages such as source for massive feedstock, relatively poor low-temperature properties, increase in NO_x emissions, etc. These issues should be sorted out before biodiesel is applied into diesel engines in a large scale. Therefore, in-depth studies on the application of biodiesel into diesel engines are necessary. The research on alternative feedstocks is also an important area and the second-generation biodiesel is more promising made from algae and the genetic modification is a potential way to solve this problem of source of massive feedstock. The low-temperature fuel properties can be improved by additives or the production routine. In addition, diesel engines should also be optimised in order to achieve the optimal performance and emissions.

Abbreviations:

ASOI	After start of injection
BTL	Biomass-to-liquid
BSFC	Brake specific fuel consumption
CCD	Charge-coupled device
CN	Cetane number
CO	Carbon monoxide
CTL	Coal-to-liquid
EPA	Environmental Protection Agency
FAME	Fatty acid methyl ester
GTL	Gas-to-liquid

HC	Hydrocarbon
HVO	Hydro-treated vegetable oil
NOx	Nitric oxide
PAH	Polycyclic aromatic hydrocarbon
RME	Rapeseed methyl ester
ULSD	Ultra-low sulphur diesel

Author details

Yanfei Li[1,2*], Guohong Tian[3] and Hongming Xu[1,2]

*Address all correspondence to: yanfei.lee@gmail.com

1 School of Mechanical Engineering, University of Birmingham, UK

2 State Key Laboratory of Automotive Safety and Energy, Tsinghua University, China

3 Sir Joseph Swan Centre for Energy Research, Newcastle University, UK

References

[1] International Energ Outlook. (2011). U.S. Energy Information Administration.

[2] Alam, M., Song, J., et al. (2004). Combustion and Emission Performance of Low Sulfur Ultra-Low Sulfur and Biodiesel Blends in a DI Diesel Engine. *SAE paper 2004-01-3024.*

[3] Allen, C. A. W., & Watts, K. C. (2000). Comparative analysis of the atomization characteristics of fifteen biodiesel fuel types. *Transactions of the Asae,* 43(2), 207-211.

[4] Allen, C. A. W., Watts, K. C., et al. (1999). Predicting the viscosity of biodiesel fuels from their fatty acid ester composition. *Fuel,* 78(11), 1319-1326.

[5] Ban-Weiss, G. A., Chen, J. Y., et al. (2007). A numerical investigation into the anomalous slight NOx increase when burning biodiesel; A new (old) theory. *Fuel Processing Technology,* 88(7), 659-667.

[6] Cheng, A., Upatnieks, A., et al. (2006). Investigation of the impact of biodiesel fuelling on NOx emissions using an optical direct injection diesel engine. *International Journal of Engine Research,* 7(4), 297-318.

[7] Chuepeng, S., Tsolakis, A., et al. (2007). A Study of quantitative impact on emissions of high proportion of RME-based biodiesel blends. *SAE technical paper 2007-01-0072.*

[8] Chuepeng, S., Xu, H. M., et al. (2008). Particulate emissions from a common rail fuel injection diesel engine with RME-based biodiesel blended fuelling using thermo-gravimetric analysis. *SAE technical paper 2008-01-0074.*

[9] Desantes, J. M., Payri, R., et al. (2009). Influence on Diesel Injection Characteristics and Behavior Using Biodiesel Fuels. *SAE 2009-01-0851.*

[10] EPA. (2002). A Comprehensive Analysis of Biodiesel Impacts on Exhaust Emissions. *Environmental Protection Agency.*

[11] EPA. (2009). Biodiesel Handling and Use Guidelines. *4th Edition, EPA. DOE/ GO-102006-356.*

[12] Gerpen, J. V. (2006). Cetane Number Testing of Biodiesel.

[13] Gong, G., Song, C., et al. (2011). Spray characteristics of diesel fuel, Fishch-Tropsch diesel fuel and their blend. *Electrical and Control Engineering (ICECE), International Conference on.*

[14] Graboski, M. S., & Mc Cormick, R. L. (1998). Combustion of fat and vegetable oil de-rived fuels in diesel engines. *Progress in Energy and Combustion Science, 24(2), 125-164.*

[15] Grimaldi, C., & Postrioti, L. (2000). Experimental Comparison Between Conventional and Bio-derived Fuel Sprays from a Common Rail Injection System . *SAE technical paper 2000-01-1252.*

[16] Knothe, G., Matheaus, A. C., et al. (2003). Cetane numbers of branched and straight-chain fatty esters determined in an ignition quality tester[small star, filled]. *Fuel, 82(8), 971-975.*

[17] Kuronen, M., Mikkonen, S., et al. (2007). Hydrotreated Vegetable Oil as Fuel for Heavy Duty Diesel Engines. *SAE technical paper 2007-01-4031.*

[18] Ladommatos, N., Abdelhalim, S. M., et al. (1998). Effects of EGR on heatrelease in diesel combustion. SAE technical paper 980184.

[19] Mc Cormick, R. L., Tennant, C. J., et al. (2005). Regulated Emissions from Biodiesel Tested in Heavy-Duty Engines Meeting 2004 Emission Standards. *SAE paper 2005-01-2200.*

[20] Mueller, C. J., Boehman, A. L., et al. (2009). An Experimental Investigation of the Ori-gin of Increased NOx Emissions When Fueling a Heavy-Duty Compression-Ignitio with Soy Biodiesel. *SAE paper 2009-01-1792.*

[21] Nguyen, D. N., Ishida, H., et al. (2009). Ignition and Combustion Characteristics of Gas-to-Liquid Fuels for Different Ambient Pressures. *Energy & Fuels, 24(1), 365-374.*

[22] Schönborn, A., Ladommatos, N., et al. (2009). The influence of molecular structure of fatty acid monoalkyl esters on diesel combustion. *Combustion and Flame,* 156(7), 1396-1412.

[23] Schmidt, K., & Gerpen, J. (1996). The effect of biodiesel fuel composition on diesel diesel combustion and emissions. *SAE paper 961086.*

[24] Senatore, A., Cardone, M., et al. (2005). Experimental Characterization of a Common Rail Engine Fuelled with Different Biodiesel. *SAE technical paper 2005-01-2207.*

[25] Sharp, C. A., Howell, S. A., et al. (2000). The effect of biodiesel fuels on transient emission from modern diesel engien~Part 1: Regulated emission and performance. *SAE paper 2000-01-1967.*

[26] Sharp, C. A., Howell, S. A., et al. (2000). The effect of biodiesel fuels on transient emissions from modern diesel engines, Part 2 unregulated emissions and chemical characterization. *SAE Transaction,* 109(4), 1784-1807.

[27] Sheehan, J., Camobreco, V., et al. (2000). An Overview of Biodiesel and Petroleum Diesel Life Cycles. Other Information: PBD: 27 Apr: Medium: ED; Size: vp.

[28] Sukjit, E., & Dearn, K. D. (2011). Enhancing the lubricity of an environmentally friendly Swedish diesel fuel MK1. *Wear,* 271(9-10), 1772-1777.

[29] Szybist, J. P., & Boehman, A. L. (2003). Behavior of a Diesel Injection System with Bi-odiesel Fuel. *SAE paper 2003-01-1039.*

[30] Terry, B., Mc Cormick, R. L., et al. (2006). Impact of Biodiesel Blends on Fuel System Component Durability. *SAE technical paper 2006-01-3279.*

[31] Wu, M., Wu, G., et al. (2005). Low-Temperature Fluidity of Bio-Diesel Fuel Prepared from Edible Vegetible Oil. *Petroleum processing and petrochemicals,* 36(4), 58-60.

[32] Zhang, J., Xu, H., et al. (2011). The Particle Emissions Characteristics of a Light Duty Diesel Engine with 10% Alternative Fuel Blends. *SAE International Journal of Fuels and Lubricant* [3], 438-452.

[33] Zhang, X., Wang, H., et al. (2008). Characteristics of Output Performances and Emissions of Diesel Engine Employed Common Rail Fueled with Biodiesel Blends from Wasted Cooking Oil. *SAE Technical Paper 2008-01-1833.*

Chapter 6

An Analysis of Physico-Chemical Properties of the Next Generation Biofuels and Their Correlation with the Requirements of Diesel Engine

Artur Malinowski, Joanna Czarnocka and
Krzysztof Biernat

Additional information is available at the end of the chapter

1. Introduction

There is a pressing need to haste developing advanced energy technologies to reduce dependency on crude oil and climate protection. Biofuels – liquid and gaseous fuels derived from organic matter – can play an important role in reducing of carbon dioxide (CO_2) emissions in the transport, and can raise the energy security. By 2050, biofuels could provide 27% of total transport fuel. The use of biofuels could avoid around 2.1 gigatonnes (Gt) of CO_2 emissions per year when produced sustainably. To meet this vision, most conventional biofuel technologies need to improve conversion efficiency, cost and overall sustainability. Conventional biofuel technologies include well-established that are already producing biofuels on a commercial scale. These biofuels, commonly referred to as first-generation, include sugar- and starch-based ethanol, oil-crop based biodiesel and straight vegetable oil, as well as biogas gained through anaerobic digestion. The International Energy Agency has undertaken an effort to develop a series of global technology road maps covering 19 advanced technologies, commonly referred to as second- or third-generation. This new technologies are still in the research and development (R&D), pilot or demonstration phase [1].Significant decrease of fossil fuels and lack of new ones becomes the basis for the Olduvai theory, published by R.C. Duncan [2]. The theory postulated that in the years 2012-2030, because of shortage of energy, the world would go through an economic crisis. This crisis would lead to collapse of industrial civilization. So, there is a need to look for alternative, renewable sources of raw materials.

According to the current EU Directive on promoting the use of energy from renewable sources [3], petrochemical companies are obliged to market fuel containing biocomponents
(2009/28/EC). Biomass means the biodegradable fraction of products, waste and residues
from biological origin, from agriculture, forestry and related industries including fisheries
and aquaculture, as well as the biodegradable fraction of industrial and municipal wastes.
Assuming biomass as the basic source of materials for the production of biofuels, two main
material pathways and the suiting material processing technologies have been considered in
the European definition. That is referred to as BtL (biomass-to-liquid) or, as an alternative,
BtG (biomass-to-gas) and WtL (waste-to-liquid) or, as an alternative, WtG (waste-to-gas).

Biofuels are divided into groups according to their state of matter. According to Annex 1 to
Communication from the Commission of the European Communities No. 34 of 2006, COM
(2006)34 final, biofuels have been divided into liquid, gas, and others, with first and second
generation biofuels having been introduced in this Communication for the first time. However, an idea of "synthetic biofuels" has been introduced and defined as "synthetic hydrocarbons or mixtures of synthetic hydrocarbons produced from biomass, for example SynGas
produced from gasification of forestry biomass or SynDiesel."

2. Classification of biofuels

In the European classification, the following biofuels have been separated because of the
state of matter:

1.   Liquid biofuels:

• Bioethanol got from biomass or biodegradable waste fractions, possible for use as biofuel
  E5 of 5% ethanol and 95% petrol contents or as biofuel E85 of 85% ethanol and 15% petrol
  contents;

• Biodiesel containing methyl-esters [PME ("pure vegetable oils"), RME ("rapeseed methyl
  esters"), FAME ("fatty acid methyl esters")] produced from vegetable oil, animal oil or recycled (for example post-frying) fats and oils, meeting the requirements of relevant quality standards for B5 diesel oils of 5% ester and 95% petroleum-based diesel contents, B30
  diesel oils of these proportions being 30% and 70%, respectively, and B100 exclusively
  consisting of pure esters of properties meeting the relevant standard specifications;

• Biomethanol produced from biomass, for use as biofuel or a fuel ingredient;

• BioETBE, that is Ethyl-tertio-butyl-ether produced from bioethanol, used as a petrol additive to increase the octane rating and to reduce knocking and added to petrol at a percentage rate of 47%;

• BioMTBE, that is Methyl-tertio-butyl-ether produced from biomethanol, used for the
  same purposes as those of the BioETBE and added to petrol at a percentage rate of 36%;

• BtL, that is Liquid fractions or mixtures of liquid fractions produced from biomass, for
  use as biofuels or fuel ingredients;

- Pure vegetable oils (PVO) produced through pressing, extraction or similar, inclusive of refining, but chemically unmodified, which can be used as biofuel when compatible with the engine involved and when meeting the matching environmental protection requirements.

2.   Gaseous biofuels:

- BioDME transport fuels gained from Renewable Energy Sources (RES), that is Dimethylether produced from biomass, for direct use as biofuel for compression-ignition engines;

- Biogas, that is Biofuel produced from biomass or the biodegradable fractions of waste, purified to natural gas quality;

- Biohydrogen as biofuel produced from biomass or the biodegradable fractions of waste.

3.   Other renewable fuels, that is Biofuels not named above, originating from sources as defined in Directive 2001/77/EC and suitable to power transport.

This division resulted from the reasons discussed above, in particular from assessment of the usability of specific fuels in the present-day engine technologies, availability of the feedstock needed, and environmental impact of the fuels. The formal division of biofuels into specific generations has been published in the report titled "Biofueels in the European Union, a Vision for 2010 and Beyond". According to this report, biofuels have been divided into first generation biofuels, referred to as "conventional biofuels," and second generation biofuels, referred to as "advanced biofuels."

The first generation ("conventional") biofuels include:

- Bioethanol (BioEtOH, BioEt), understood as conventional ethanol got through hydrolysis and fermentation from raw materials such as cereals, sugar beets.;

- Pure vegetable oils, got through cold pressing and extraction from seeds of oil plants;

- Biodiesel, consisting of RME or FAME and fatty acid ethyl esters (FAEE) of higher fatty acids of other oily plants and gained as the result of cold pressing, extraction and transesterification;

- Biodiesel, consisting of methyl and ethyl esters and gained as the result of transesterification of post-frying oil;

- Biogas, got by purification of wet landfill or agricultural biogas;

- BioETBE, got by chemical processing of bioethanol.

The idea of second generation biofuels development is based on an assumption. Feedstock to be used for producing such fuels should equally include biomass, waste vegetable oils and animal fats, as well as any waste substances of organic origin that are useless in the food and forestry industries. The second generation ("advanced") biofuels includes:

- Bioethanol, biobutanol, and blends of higher alcohols and derivative compounds, got as the result of advanced of hydrolysis and fermentation of lignocellulosic biomass (excluding the feedstock for food production purposes);

- Synthetic biofuels, being products of biomass processing and gained by gasification and proper synthesis into liquid fuel ingredients (BtL) and products of process biodegradable industrial and municipal wastes, including carbon dioxide (WtL);

- Fuels for compression-ignition engines, got from biomass through Fischer-Tropsch, inclusive of synthetic biodiesels got by blending of lignocellulosic products;

- Biomethanol, got as the result of lignocellulose transformation, inclusive of Fischer-Tropsch synthesis, as well as with the use of waste carbon dioxide;

- Biodimethylether (bioDME), got by thermochemical processing of biomass, inclusive of biomethanol, biogas, and synthetic biogases being derivative products of biomass transformation;

- Biodiesel as biofuel or a fuel ingredient for compression-ignition engines, got by hydrorefining (hydrogenation) of vegetable oils and animal fats;

- Biodimethylfuran (bioDMF), obtained from sugar transformation, inclusive of transforming cellulose in to thermochemical and biochemical processes;

- Biogas as synthetic natural gas (SNG) or biomethane, obtained in result of lignocelluloses gasification, correct synthesis, or purification of agricultural, landfill, and sewage sludge biogas;

- Biohydrogen got in result of gasification of lignocellulose and synthesis of the gasification products or as the result of biochemical processes.

The European Commission Directorate-General for Energy and Transport proposed to separate third generation biofuels, defining them as those for which the technology of universal gain and introduction of such fuels may be developed in 2030s or even later, according to the estimates. Preliminarily, biohydrogen and biomethanol have been classified in this group. The third generation biofuels may be obtained by the methods similar to those used in the second generation biofuels, but from the feedstock (biomass) having been modified at the plant growing stage with the use of molecular biology techniques. The objective of such changes is to improve the conversion of biomass into biofuels (biohydrogen, biomethanol, biobutanol) by for example cultivation of trees of low lignin content, development of crops with enzymes incorporated as required, etc.

Separating a new, fourth generation of biofuels was proposed because of the need to close the carbon dioxide balance or to cut out the environmental impact of this compound. Therefore, the fourth generation biofuel technologies should be developed with considering the CCS ("Carbon Capture and Storage") at the raw material preparation and biofuel production stages. The raw materials used for production of such fuels should be the plants of increased CO_2 assimilation rates at the plant growing stage and the technologies applied must be devised considering the capture of carbon dioxide in proper geological formations by causing the carbonate stage to be reached or the storage in oil and gas exploitation cages.

3. The main directions of advanced fuel technology's development

Within the planned perspective of the production and use of biofuels, the fuels are required: to be available in enough large quantities; to have acceptable technical and energy character-istics for being suitable for fueling engines or heating; to be inexpensive at both the produc-tion and sale stages; to cause smaller environmental hazard in comparison with the conventional fuels; to improve energy independence.

Based on the experience and on results of the research work carried out, we should strive in the nearest future to get biofuels as hydrocarbon blends produced by definite pathways. Such pathways will make it possible to get alternative fuels for IC engines with simultane-ous closing of the CO_2 cycle. Therefore, the advanced biofuels should be:

- Synthetic biofuels made as blends of hydrocarbons produced in result of biomass gasifica-tion and pyrolysis [4] (figures 1 and 2)

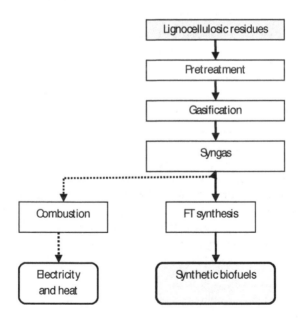

Figure 1. Schematic diagram of biomass to liquid process in Choren, Germany.

The main piece of biomass gasification technology is the patented Carbo-V process that al-lows to produce tar-free synthetic gas, a breakthrough for biomass to energy change. The gas consisting mainly of CO and H_2 can be used as a combustion gas for the generation of electricity, steam or heat, or for the make of transport fuels (BtL). Compared with fossil diesel, the combustion of BtL diesel reduces PM's (particulate matters) emissions by 30 to 50% and hydrocarbon emissions by up to 90 %. It achieves superior combustion characteristics while no engine adjustments are needed. But perhaps its most important feature is the ability to recycle atmospheric CO_2 into the fuel thus closing the sustainability cycle.

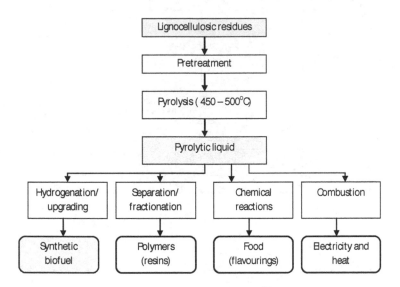

Figure 2. Schematic diagram of Rapid Thermal Processing (RTP) TM technology in Ontario, Canada (owner Ensyn). It is commercial installation.

Another promising technology is related to pyrolysis. Rapid Thermal Processing (RTP) is a fast thermal process where biomass is rapidly heated without oxygen. The biomass is vaporized and rapidly cooled to produce high yields of pyrolysis oil. The pyrolysis oil is fractionated into chemicals for engineered wood products (resins) and fuel for thermal applications. The resulting char and gases are used for energy. RTP™ typically yields 65 to 75wt% pyrolysis oil from dried woody biomass.

• Biofuels earned from biomass in result of other thermochemical processes, such as pyrolysis or processes of depolymerisation and hydrogenation of biomass decomposition products (hydrothermal upgrading-HTU processes);

• Fuel blends composed of hydrocarbons gained from biomass, including those directly or indirectly obtained from sugars in result of biological or chemical processes,

• Biofuels being other sugar derivatives;

• Biomethane and other gaseous fuels got from biomass gasification processes or agricultural, landfill, and sewage sludge treatment processes;

• Bioethanol and higher alcohols -biobutanol and their derivatives, obtained from biomass
  in result of biochemical or catalyzed thermochemical processes (figure 3);

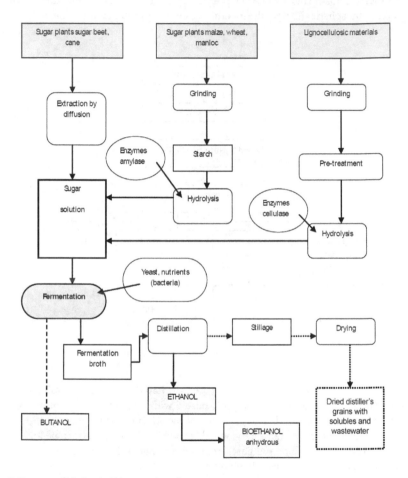

Figure 3. Processes of biochemical biomass conversion.

• Biofuels obtained by utilization of carbon dioxide for production of microorganisms or by
  direct or indirect synthesis of carbon dioxide of natural origin in thermochemical and biochemical;

• Biofuels obtained from synthetic gas produced as a product of direct or indirect (through
  methanol) conversion of biomass or GHG;

• Biofuels (HVO, hydrogenated vegetable oils) got by hydrogenation of waste vegetable
  and animal fats.

4. Application of selected types of biofuels of the first and second generation

Among the proposed alternative fuels, vegetable oils have received much attention in recent years for diesel engines owing to their advantages as renewable and domestically produced energy. The major disadvantage of pure vegetable oils is their inherently high viscosity, leading to poor fuel atomization, incomplete combustion, coking of fuel injectors, ring carbonization, and accumulation of vegetable oil in the lubricating oil. Several methods are consequently being used to reduce vegetable oil's viscosity. Blending of vegetable oils with an alcohol of lower viscosity is one of the methods [5, 6].

Main alcohols used as a fuels ingredient are: methanol, ethanol and n-butanol. These alcohols have different properties. Some of them are presented in Table 1. They are compared to conventional engine fuels.

Fuel	Energy density	Heat of vaporization	Kinematic viscosity at 20°C
Diesel	38.6 MJ/l	0.47 MJ/kg	"/>3 cSt
Gasoline	32.0 MJ/l	0.36 MJ/kg	0.4–0.8 cSt
Butanol	29.2 MJ/l	0.43 MJ/kg	3.64 cSt
Ethanol	19.6 MJ/l	0.92 MJ/kg	1.52 cSt
Methanol	16.0 MJ/l	1.20 MJ/kg	0.64 cSt

Table 1. The properties of different alcohols and engine fuels

It is interesting the butanol has similar energy density as petrol. Butanol is good solvent of heavy hydrocarbons (such diesel fuels). The mixture of these components is homogeneous and doesn't separate after several months. In contrast, ethanol is slightly soluble in diesel fuel. It is important the water is nearly insoluble in butanol, in contrast to ethanol which dissolves water in any proportion.

The old and new technology of butanol production is known as an ABE process (Acetone-Butanol-Ethanol) and the second generation process using lignocellulosic waste materials, respectively. The conventional ABE fermentation process is based on sugar's material (cane or beet) or starch (wheat, corn or rice) which is easily broken down into sugars. During the fermentation formed the three components: acetone, n-butanol and ethanol (in ratio of 3:6:1) as main products. The process is performed by anaerobic gram- positive bacteria of the geneus clostridia (mainly Clostridium acetobutylicum, but also C. Beijerinckii, C. butylicum and others). The ABE process is not profitable because of low productivity and poor selectivity. One of the courses covers metabolic engineering issues, that is modification of metabolic pathway to increase resisting clostridia bacteria to higher concentrations of fermentation products, and improve the efficiency and selectivity. Low yield of the fermentation of butanol synthesis requires research on butanol recovery techniques. There are many separation techniques of fermentation products, e. g. liquid-liquid extraction, perstraction, pervaporation (membrane separation with gaseous permeate discharge) combined with immobilization of bacterial cells, adsorption or reversed osmosis. It is estimated that effective solutions development can help to increase of profitability up to 40-50%.

The ideal feedstock for bioconversions could be waste biomass, for example straw, wood chips and paper pulp effluent. Also crops specially grown for their high biomass production rate (kenaf, miscanthus and short rotation woody crops). Such sources of raw materials can be described as "cellulosic biomass" because of their high cellulose and hemicellulose content. The feedstock used in fermentation determines the selection of strains and process conditions. The company Green Biologics is developing biobutanol production from glycerol and other wastes from industry and agriculture, using genetically modified thermophilic bacteria of the genus geobacillus and sells derived fuel named Butafuel (figure 4).

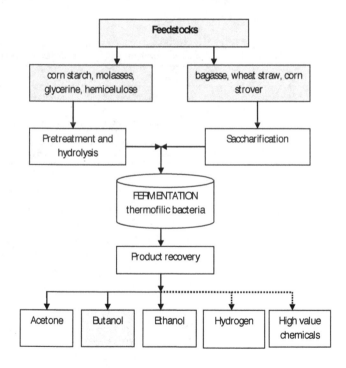

Figure 4. Schematic diagram of GBL's technology, Green Biologics Ltd. Biobutanol used to produce fuels, paints, coatings, resins, polymers and solvents.

Butanol, like ethanol, can blend well with gasoline. Biobutanol can replace gasoline in E85 fuel. Also, butanol could be a future for blending with diesel. Butanol contains more oxygen compared to the biodiesel, leading to further decline of soot. NO_x emissions can also be reduced because of its higher heat of evaporation, which results in a lower combustion temperature. The butanol has more advantages than the widely used ethanol and FAME. However, the main disadvantage of butanol is low production. Biobutanol is noncorrosive and can be shipped via pipeline.

No.	Butanol isomers	Main application
1	1-butanol	Gasoline additive, solvents, plasticizers, chemical intermediate, cosmetics
2	2-butanol	Solvents, chemical intermediate, industrial cleaners, perfumes or in artificial flavors
3	iso-butanol	Gasoline additive, solvent and additive for paint, industrial cleaners, ink ingredient
4	tert-butanol	Gasoline additive for octane booster and oxygenate; intermediate for MTBE, ETBE, THBP; denaturant for ethanol; solvent

Table 2. The main application of butanol isomers. Ref. [7]

This work presents a novel way of using alcohols and pure vegetable oil as fuels for a diesel engine. It was shown the possibility of use of higher alcohols as a solvent for straight vegetable oil (the mixture was named BM). Such a mixture, after getting the density similar to the density of diesel fuel, was mixed with diesel fuel (D) giving biomixdiesel (BMD). For BMD preparation was used the n-butanol and iso-amyl (by-product of ethanol fermentation) as an alcohol, rapeseed oil and conventional diesel fuel. Another biofuels as an example of second generation were obtained by nonoxidative thermal/pyrolytic cracking of straw (nearly 200 microns) followed by biooil hydrotreating. The last one- HVO diesel was obtained by catalytic hydroconversion of vegetable oil mixtures. Hydrotreated vegetable oils do not have the harmful effects of ester-type biodiesel fuels, like increased NO_x emission, deposit formation, storage stability problems, more rapid aging of engine oil or poor cold properties. HVOs are straight chain paraffinic hydrocarbons that are free of aromatics, oxygen and sulfur and have high cetane numbers. All three biofuels were examined according to EN-590:2009 standard [8]

4.1. Experimental and results

4.1.1. Assessment of the physico-chemical properties of the BMD biofuel

To assess the quality of biofuel containing components such as higher alcohol and rapeseed oil were prepared two experimental blends based on previous works [9, 10]. Major scientific works regards diesel-biobutanol mixtures [11] and minor triple mixtures with vegetable oil. The main component of mixtures was conventional diesel (in 80% vol.), made up to 100% with two mentioned above biocomponets. Experiment was carried out with two higher alcohols: n-butyl alcohol and iso-amyl alcohol. First were prepared blends consisting of selected alcohol and rapeseed oil in a ratio of 2:1 (BioMix), and then received blend was introduced into diesel fuel (D). Prepared samples are marked with symbols BMD-1 (with n-butanol) and BMD-2 (with iso-amyl alcohol). Mixtures of BMD-1 and BMD-2 were clear, without haze and sediment. New biofuels stored for several days at room temperature showed no features of separation. Diesel fuel used to compose biofuels met all quality requirements according to EN-590. Table 3 shows the basic features of diesel, and Table 4 compares properties of n-butyl alcohol and iso-amyl.

No.	Property	Result
1.	Cetane number	53,0
2.	Density at 15°C, kg/m³	836,2
3.	Flash point, °C	63
4.	Carbon residue (on 10% distillation residue), %(m/m)	<0,10
5.	Distillation %(V/V) recovered at 250°C, %(V/V) recovered at 350°C, 50%(V/V) recovered at , °C 95%(V/V) recovered at , °C finish boiling point, °C	39,5 94,9 266,7 350,5 362,4

Table 3. Basic physico-chemical properties of diesel fuel

No.	Property	n-butyl alcohol	iso-amyl alcohol
1.	Density at 20 °C, kg/m³	810	814
2.	Boiling point °C	117	138
3.	Flash point, °C	30	43

Table 4. Properties of n-butyl alcohol and iso-amyl alcohol

Prepared biofuels samples were examined according with regulatory needs of the standard EN 590. The results got are presented in Table 5 and Table 6.

No.	Property	Test method	Result BMD-1	Result BMD-2	Limits EN 590
1	Cetane number	EN ISO 5165	44,4	45,0	min 51,0
2	Cetane index	EN ISO 4264	46,8	46,9	min 46,0
3	Density at 15°C, kg/m³	EN ISO 12185	837,9	837,8	820,0 – 845,0
4	Polycyclic aromatic hydrocarbons, % (m/m)	EN 12916	1,9	1,9	max 11
5	Sulfur content, mg/kg	EN ISO 20846	5,7	5,7	max 10,0
6	Flash point, °C	EN ISO 2719	< 40,0	45,0	above 55
7	Carbon residue (on 10% distillation residue), %(m/m)	EN ISO 10370	0,48	0,27	max 0,30
8	Ash content, %(m/m)	EN ISO 6245	< 0,001	< 0,001	max 0,01
9	Water content, mg/kg	EN ISO 12937	110	110	max 200
10	Total contamination, mg/kg	EN 12662	<6,0	9,0	max 24
11	Copper strip corrosion (3 h at 50°C)	EN ISO 2160	class 1	class 1	class 1
12	Lubricity, corrected wear scar diameter (wsd 1,4) at 60°C , µm	EN ISO 12156-1	281	339	max 460
13	Viscosity at 40°C, mm²/s	EN ISO 3104	2,710	2,827	2,00 – 4,50
14	Distillation %(V/V) recovered at 250°C, %(V/V) recovered at 350°C, 50%(V/V) recovered at , °C 95%(V/V) recovered at , °C Finish boiling point, °C	EN ISO 3405	47,3 261,9 349,9	44,2 261,4 349,7	< 65 min 85 max 360
15	Fatty acid methyl ester content, FAME , %(V/V)	EN 14078	< 1,6	< 1,6	max 7,0
16	Oxidation stability, g/m³	EN-ISO 12205	66	39	max 25

Table 5. Comparison of the results of biofuels BMD-1 and BMD-2 according to EN 590

No.	Property	Test method	Result BMD-1	Result BMD-2	Summer	Spring and autumn	Winter
1	Cold filter plugging point, CFPP, °C	EN 116	-21	-21	max 0	max -10	max -20
2	Cloud point, °C	ISO 3015	-6	-6	limits only for arctic climate		

Table 6. Comparison of low-temperature properties of fuels BMD-1 and BMD-2 to the climatic requirements of EN 590

Comparing the results of biofuels BMD-1 and BMD-2 with quality requirements for diesel fuel it is worth to note that most of the parameters meet these requirements; however, several features deviate from the normative requirements. Cetane number is similar on both biofuels and amounted 44.4 and 45.0 for BMD-1 and BMD-2, respectively and is lower than the required standard that is at least 51 units. This is because of sharing 20% biocomponents. Rapeseed oil has a cetane number about of 40-50 units and a small addition to the diesel fuel should not drastically reduce the cetane number. However alcohol is usually characterized by a high octane number, which is good in case of composing gasoline, added to the diesel fuel can degrade the diesel engine start-up parameters.

The process of starting engine and his operation is also influenced by fractional composition of fuel, particularly temperature distillation of 50% by volume of fuel, T_{50}. The lower the temperature T_{50} the easier the start, but at too low temperature ignition characteristics fuel property is worsen - cetane number decreases.

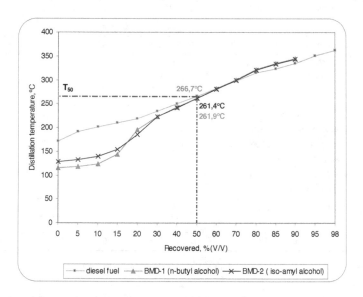

Figure 5. Distillation of diesel fuel, biofuels BMD-1 and BMD-2 comparison.

Figure 5 shows the distillation composed of biofuels compared to diesel. T_{50} temperatures for the tested biofuels BMD-1 and BMD-2 are similar and amounted 261.9 and 261.4°C, respectively and slightly differs from the T_{50} for diesel fuel - 266.7 °C. The temperature range from about 250 °C to 340 °C showed the curves of the distillation of biofuels and diesel fuels are similar. The beginning of the distillation is unusual for sample BMD-1, which begins to boil in temperature 114°C, and BMD-2 at a temperature of 127°C. Alcohol is distilled off first at the early stage of the distillation followed by hydrocarbons and rapeseed oil. Distillation points out the biofuel combustion in the engine can be irregular.

Contributing alcohol in biofuel can cause the decline of ignition temperature. Flash point of biofuel blend is determined by its flash point of the lightest ingredient, which n-butyl alcohol and iso-amyl alcohol was 37 °C, and 45 °C, respectively. Safety in transport and storage of diesel fuel requires the ignition temperature be higher than 55°C. Thus, the fuel with a lower flash point cannot be marketed and sold, could possibly be used as fuel for selected fleets.

The tested biofuel has a high tendency to form sludge and carbon deposits, which is determined by the remains of carbon residues. BMD-1 sample carbon residue value exceeds regulatory requirements by 60%, for the BMD-2 approaches the limit. Biofuel with such high carbon residues will cause form deposits in the combustion chamber, the valves, piston rings and injector parts. Sediments and carbon residues can change conditions of heat exchange, worsen the quality of fuel atomization and eventually can lead to immobilization of the vehicle.

Both samples of biofuels have good low temperature properties (Table 6). Tested cold filter blocking temperature for both samples amounted -21 °C, and the cloud temperature -6°C and slightly was different from the similar parameters of diesel. Low-temperature stability studies have shown that biofuel stored for several days at a temperature of about-10°C becomes cloudy, but segregation was not observed and had the liquid properties. Viscosity of biofuels is correct and for BMD-1 and BMD-2 were 2.710 mm²/s and 2.827 mm²/s, respectively. Proper fuel viscosity is important, because directly influence on the quality of atomization and combustion. Other biofuels quality parameters measured; do not differ from the normative requirements. The sulfur and water content, polycyclic aromatic hydrocarbons, solids, ash residue, lubricity and density are within the limits. Please note that this biofuel may not be used for long-term storage, it is unsatisfactory because of their oxidation stability. The sludge after being marked with an accelerated aging process is large, about 2 times the standard requirements. It is therefore recommended product produced in small quantities, intended for fast using.

Tested biofuels BMD-1 and BMD-2 were assessed for regulatory quality requirements. It is difficult to clearly settle, which biofuel blends is better. Features such as low cetane number, low flash point, and atypical distillation limit the usefulness of both biofuels to power the diesel engines. Preliminary experiments should be continued for improvement to compose biofuels and carry out the procedure for selecting additives. It is necessary to increase the cetane number. The proper corrosion protection should be considered because of the presence of alcohol ingredient in biofuel. To introduce new BMD biofuels still needs much research and formula improvements.

4.1.2. Emission test for BMD fuel

Based on the physicochemical properties of biofuels BMD-1 and-2 for further test was selected BMD mixture with n-butanol only. Prepared under laboratory conditions mixtures of n-butanol with diesel fuel were examined on the chassis dynamometer. In the first step the rapeseed oil was mixed with butanol as such parts to obtain a mixture having a density similar to the density of diesel fuel. This mixture is denoted as a BM (BioMix). In the second step this fuel (BM) was mixed with conventional diesel fuel (D) to get biomixdiesel (marked as a BMD). These fluids were mixed in the following parts:

- biomix (BM) 20 % v/v,

- diesel fuel (D) 80 % v/v,

giving fuel called as biomixdiesel (BMD20). In contrast to the mixture of ethanol with rape methyl ester and conventional diesel fuel, this mixture is homogeneous. The comparison of new fuel with requirements of the standard diesel fuel is presented in Table 7.

No.	Property	Test method	Results BMD20	Limits EN 590
1	Cetane number	EN ISO 5165	44,4	min 51,0
2	Cetane index	EN ISO 4264	46,8	min 46,0
3	Density at 15ºC, kg/m³	EN ISO 12185	837,3	820,0 – 845,0
4	Polycyclic aromatic hydrocarbons, %(m/m)	EN 12916	1,9	max 11
5	Sulfur content, mg/kg	EN ISO 20846	5,7	max 10,0
6	Flash point, ºC	EN ISO 2719	< 40,0	above 55
7	Carbon residue (on 10% distillation residue), % (m/m)	EN ISO 10370	0,48	max 0,30
8	Ash content, %(m/m)	EN ISO 6245	< 0,001	max 0,01
9	Water content, mg/kg	EN ISO 12937	112	max 200
10	Total contamination, mg/kg	EN 12662	<6,0	max 24
11	Copper strip corrosion (3 h at 50ºC)	EN ISO 2160	class 1	class 1
12	Lubricity, corrected wear scar diameter (wsd 1,4) at 60°C , µm	EN ISO 12156-1	281	max 460
13	Viscosity at 40ºC, mm²/s	EN ISO 3104	2,710	2,00 – 4,50
14	Distillation %(V/V) recovered at 250ºC, %(V/V) recovered at 350ºC, 95%(V/V) recovered at , ºC Finish boiling point, ºC	EN ISO 3405	47,3 349,9	< 65 min 85 max 360
15	Fatty acid methyl ester content, FAME , %(V/V)	EN 14078	< 1,6	max 7,0
16	Oxidation stability, g/m3	ISO 12205	66	max 25

Table 7. Properties of investigated fuel

The investigations of fuel properties under working conditions were carried out with a modern diesel engine on the chassis bed dynanomyter in the NEDC test (New European Driving Cycle). This test consists of two parts: UDC (Urban Driving Cycle) and EUDC (Extra Urban Driving Cycle). The first part represents urban driving, in which a vehicle is started in the morning (after being parked all-night) and driven in stop-and-go rush hour traffic. The maximum speed is 50 km/h. The second part represents extra-urban driving with a maximum speed of 120 km/h.

Main parameters of car engine (power, torque, specific fuel consumption) and the main exhaust gas ingredient (in this case CO, CO_2, NO_x, total hydrocarbons-THC, particulate matter-PM, THC+ NO_x) and fuel consumption is evaluated and explain here in g/km. In Table 8 obtained are presented results.

Emission	Fuel	Pollutants, g/km						Fuel consumption
		THC	CO	CO_2	NO_x	THC+NO_x	PM	g/km
UDC	BMD 20	0,1000	1,3900	163,6367	0,1933	0,2967	0,0042	6,2467
UDC	Diesel	0,0833	1,3400	162,1367	0,2000	0,2867	0,0053	6,1933
EUDC	BMD 20	0,0100	0,0367	118,7567	0,2133	0,2267	0,0064	4,4667
EUDC	Diesel	0,0100	0,0467	114,5500	0,1600	0,1700	0,0078	4,3167
NEDC	BMD 20	0,0467	0,5367	135,2933	0,2067	0,2533	0,0056	5,1233
NEDC	Diesel	0,0367	0,5233	132,1100	0,1767	0,2167	0,0069	5,0100

Table 8. Examples of investigations results on the car test chassis bed by NEDC test load and by fueling the engine with examined BMD20 and standard Diesel fuel

Supplying the car engine with different fuel leads to a diversity of parameters of the engine. But the differences are not so significant. Differences between the results got for the tested BMD20 fuel and diesel fuel are presented below. The results of investigations of pollutant emission are presented here as the results got by fueling the engine with the conventional diesel fuel. The results expressed in g/km are shown in Table 9.

Test	Pollutants						Fuel consumption
	THC	CO	CO_2	NO_x	THC+NO_x	PM	
				g/km			
UDC	0,0167	0,0500	1,5000	-0,0067	0,0100	-0,0011	0,0533
EUDC	0,0000	-0,0100	4,2067	0,0533	0,0567	-0,0014	0,1500
NEDC	0,0100	0,0133	3,1833	0,0300	0,0367	-0,0013	0,1133

Table 9. Relative changes of pollutants emission and fuel consumption by fueling the car engine with BMD20 and Diesel fuel

The results are presented in graphical form as well on the Figure 6.

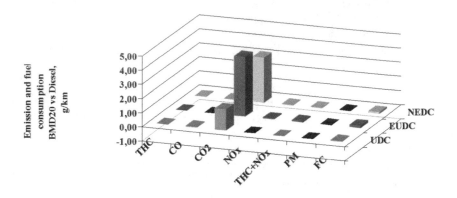

Pollutants and fuel consumption (FC)

Figure 6. The differences in emission and fuel consumption during fueling the car engine with BMD20 and Diesel fuel recorded during the test bed investigation of the car

In all phases of the test, during fueling the engine with BMD20, the increase of the carbon dioxide (CO_2) is observed. Emissions of other toxic components and fuel consumption don't differ from those when the engine is fueled by conventional fuel.

Small differences in the results are becoming clear when relative changes are expressed in percentages. The results of this analysis are presented in Table 10 and pictured on Figure 6.

Test	Pollutants						Fuel consumption
	THC	CO	CO_2	NO_x	THC+NO_x	PM	
	%						
UDC	20,00	3,73	0,93	-3,33	3,49	-20,75	0,86
EUDC	0,00	-21,43	3,67	33,33	33,33	-18,38	3,47
NEDC	27,27	2,55	2,41	16,98	16,92	-18,93	2,26

Table 10. Relative changes of pollutants emission and fuel consumption during engine fueling with BMD20 and Diesel fuel

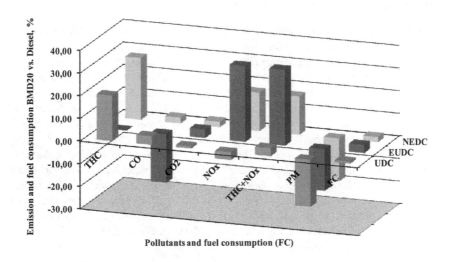

Figure 7. Relative changes of emissions and fuel consumption of the car engine fueled with BMD20 and diesel oil.

During fueling the car engine with the BMD20, the fuel consumption is not significantly different (Table 10) from that noted for Diesel fuels. The differences in emitting pollutants were dependant on the test phase (UDC or EUDC). For example the emission of THC in the EUDC phase is the same as that recorded for diesel oil, but the quantity of THC grows in UDC phase. Thus in full test (NEDC) the relative emission of THC grows. Other trend is observed for carbon monoxide (CO) emissions. In the UDC phase emission of CO slightly increases, in the EUDC phase significantly decreases (more then 21%) so, therefore in the NEDC test the emission of CO slightly grows. The emissions of NO_x grows, first, in the EUDC phase. This is understandable if we take into consideration the engine load – the higher combustion temperatures (peaks) in this phase favor to form nitrogen oxides. In the same phase the emission of THC neither increases nor decreases, so in the entire test, the summary quantity of $THC+NO_x$ increases (the emission of THC increases in the UDC phase). It is important that, the PM emission decreases in all phases of test. The decrease is significant, about 21% in the UDC phase and more than 18% in the EUDC phase.

Results were obtained without any change of engine control parameters (the engine control parameters were the same as during supplying the engine with conventional diesel fuel). It seems that after optimization of engine control features the results would be much better.

4.2. Assessment of the physico-chemical properties of the synthetic diesel fuel obtained by biomass depolymerization

Biomass depolymerization is a process for the reduction of complex organic materials. Under pressure and heat, long chain polymers of hydrogen, oxygen, and carbon decompose into biocrude oil. The depolymerization process for fuel production from organic materials takes two forms, thermal only or with assisted catalysts usually aluminum silicate type doped with non-precious metals for example Na, Ca. Although the thermal depolymerization has been understood for some time, human-designed processes were not efficient enough to serve as a practical source of fuel because more energy was required than was produced. Research breakthroughs in the 1980's led to efficient processes that were eventually commercialized [12, 13]. Green diesel was obtained by non-oxidative thermal/pyrolytic cracking of straw (around 200 micron) followed by biooil upgrading in hydrogen process. Fuel had a clear yellow color and has been tested in accordance with EN 590, one parameter not included in the specification standard (iodine number) was checked because of some doubts about non-saturated hydrocarbons content. It's known that if these hydrocarbons are present in significant quantity in the fuel, than it will cause polymerization and change the physico-chemical properties of the fuel.

No.	Property	Test method	Results	Limits EN 590
1	Cetane number	EN ISO 5165	58,6	min 51,0
2	Density at 15°C, kg/m³	EN ISO 12185	815	820,0 – 845,0
3	Polycyclic aromatic hydrocarbons, %(m/m)	EN 12916	0,8	max 11
4	Sulfur content, mg/kg	EN ISO 20846	120	max 10,0
5	Flash point, °C	EN ISO 2719	43,5	above 55
6	Water content, mg/kg	EN ISO 12937	150	max 200
7	Viscosity at 40°C, mm²/s	EN ISO 3104	3,45	2,00 – 4,50
8	Iodine number , gI/100g		8	

Table 11. The results of synthetic biofuel (Green diesel) according to EN 590.

Discussion of above parameters in Table 11 are presented below

- Cetane number - the standard EN 590 required a minimum cetane number- 51, fuels with higher cetane number have shorter ignition delays, providing more time for the fuel combustion process to be completed. Generally the engine is "soft", it's easier on the speed falls, helps to start the engine, slows pollution injector nozzles, limits the participation of PM in the exhausted gas and reduces engine noise. In the tested fuel cetane number was 58.8.

- The content of polycyclic aromatic hydrocarbons (PAHs) - in the test fuel PAH content amounted 0.8 and was significantly below the maximum value 11

- Density at 15 °C - has been marked below the lower limit of the density of 820 kg/m³ and amounted 818 kg/m³ for tested fuel. It could be adjusted by distillation the light fraction.

- Water content - the test fuel contained 170 mg/kg of the water and the result was below the maximum value of 200 mg/kg.

- Flash point - It was found the ignition temperature of the test fuel is below the value of 55 °C and was 43.5 °C, which was caused by the presence of a light fraction of the fuel.

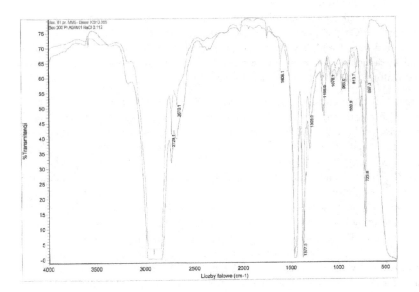

Figure 8. The comparison of IR spectrums of commercial diesel oil (red spectrum) and a sample of the green diesel (blue spectrum).

The other limits of EN 590 were not tested as the fuel is still under technological development but above results are promising. Sulfur content in fuel after preliminary hydrodesulphurization process significantly exceeded specification value of 10 ppm Infrared spectrum of green diesel was compared with spectrum of commercial diesel oil (Figure 8).

IR spectrums of analyzed commercial diesel and green diesel were similar in appearance, what shows that both fuels contain similar chemical groups. The spectrums show little difference, that reveals the presence in the spectrum of green diesel peaks at wave number 1606 cm^{-1} and around 814 cm^{-1}. These bands can be attributed to the unsaturated bonding (alkenes, aromatics). The spectrum has also characteristic bands of the groups - CH$_3$ and-CH$_2$ alkanes, including alkanes having carbon numbers greater than four, wave number band around 723 cm^{-1}. The lower band intensity in comparison with the spectrum of commercial diesel may point out a lower content of long-chain hydrocarbon sample. Analyzes mark the tested green diesel could meet the needs for diesel EN 590 if the level of sulfur below 10 ppm, increased the density and required temperature of ignition would be achieved.

4.3. Assessment of the physico-chemical properties of the synthetic diesel fuel – HVO diesel

Hydrotreating of vegetable oils or animal fats is an alternative to esterification for producing FAME. Hydrotreated vegetable oils do not have the harmful effects of ester-type biodiesel fuels, like increased NO$_x$ emission, storage stability problems and poor cold properties. HVOs are straight chain paraffinic hydrocarbons that are free of aromatics, oxygen and sulfur and have high cetane numbers. Examined HVO diesel oil was obtained at demonstration plant by catalytic hydroconversion of straight rapeseed oil. Selected features of HVO diesel are presented in Table 12.

Characteristics presented in Table 12 confirmed superior properties of HVO diesel compared with commercial diesel oil. It does not contain polynuclear aromatic and has low sulfur content. HVO diesel does not exceed the limit values for water. This product has a low tendency to foam, test parameters meet the requirements of EN 590 grade F. This means HVO diesel itself, without additives has excellent winter properties. Also the HVO diesel had a high resistance to oxidation which 2.5 times exceeds the needed normative values. During the HFRR test (High Frequency Reciprocating Rig, not presented here) was indicated the lubricant film produced by the sample is unstable, which is obvious at the early stage of the test. Typical diesel oil film thickness should exceed 90%, while in the HVO diesel case was only 78%. This shows the inadequate lubricating properties of the sample and that a lubricating additive is necessary. HVO is based on the hydrotreating, which could be used not only to convert plant-derived oils such as soybean, rapeseed, and palm, but also non-edible oils, such as Jatropha and algal oils as well as animal fats [14]. UOP and Eni have developed the Ecofining ™ based on conventional hydroprocessing technology to produce diesel-fuel ("green diesel") [15] or jet-fuel [16]. Similar technologies have been developed by Neste Oil (NExBTL Renewable Diesel) [17] and Petrobras (H-Bio process) [18].

No.	Property	Test method	Results	Limits EN 590
1	Cetane number	EN ISO 5165	"/> 76	min 51,0
2	Cetane index	EN ISO 4264	91,8	min 46,0
3	Density at 15°C, kg/m3	EN ISO 12185	776,5	820,0 – 845,0
4	Polycyclic aromatic hydrocarbons,	EN 12916		max 11
	mono, %(m/m)		< 6,0	
	di+, %(m/m)		< 1,0	
	tri+, %(m/m)		< 0,1	
5	Sulfur content, mg/kg	EN ISO 20846	< 3	max 10,0
6	Flash point, °C	EN ISO 2719	71	above 55
7	Carbon residue (on 10% distillation residue), %(m/m)	EN ISO 10370	< 0,01	max 0,30
8	Ash content, %(m/m)	EN ISO 6245	< 0,001	max 0,01
9	Water content, mg/kg	EN ISO 12937	20	max 200
10	Total contamination, mg/kg	EN 12662	1	max 24
11	Copper strip corrosion (3 h at 500C)	EN ISO 2160	class 1	class 1
12	Lubricity, corrected wear scar diameter (wsd 1,4) at 60°C , μm	EN ISO 12156-1	313	max 460
13	Viscosity at 40°C, mm2/s	EN ISO 3104	2,609	2,00 – 4,50
14	Viscosity at 50°C, mm2/s	EN ISO 3104	2,183	---
15	Distillation	EN ISO 3405		
	Initial boiling point, oC		190,4	--
	5 %(V/V) recovered at , oC		225,4	--
	10 %(V/V) recovered at , oC		242,6	--
	20 %(V/V) recovered at , oC		259,0	--
	30 %(V/V) recovered at , oC		267,3	--
	40 %(V/V) recovered at , oC		271,8	--
	50 %(V/V) recovered at , oC		274,9	--
	60 %(V/V) recovered at , oC		277,6	--
	70 %(V/V) recovered at , oC		280,3	--
	80 %(V/V) recovered at , oC		283,8	--
	90 %(V/V) recovered at , oC		288,2	--
	95 %(V/V) recovered at , oC		291,9	--
	Finish boiling point, oC		306,1	max 360
	Recovery, %(V/V)		98,3	--
	Residue, %(V/V)		1,2	--
	%(V/V) recovered at 250oC		13,5	< 65
16	CFPP, oC	EN 116	-19	*
17	Cloud point, oC	ISO 3015	-18	^
18	Pour point, oC	ISO 3016	-21	--
19	Oxidation stability, g/m3	ISO 12205	10,5	max 25

No.	Property	Test method	Results	Limits EN 590
20	Heating value - lower, MJ/kg - higher, MJ/kg	PN-C-04062	43,7 46,8	-- --
21	Composition: - hydrogen, %(m/m) - carbon, %(m/m) - nitrogen, %(m/m)	ASTM D5291	14,7 85,5 0,1	-- -- --

*- depending on the climatic requirements

Table 12. Comparison of the results of synthetic biofuel (HVO diesel) with EN 590.

5. Conclusions

Blends of n-butanol, rapeseed oil and conventional diesel fuel showed promising results and what is most important the biofuel was prepared by simple blending of biocomponents with diesel oil. Nearly the same fuel consumption compared with diesel oil was noted and the emissions of main toxic compounds including PM decreased. The other two second generation Green and HVO goals met main requirements of standard specification EN 590. The use of biofuels in transport depends on few causes like: availability of raw materials, low cost production of biofuels, low selling price, the calorific value, high quality and compliance with the needs of fuels for automotive engines.

Synthetic hydrocarbon fuels, considered as the best solution replacements of fossil fuels, may be obtained by biomass gasification followed by FT process, biomass pyrolysis towards biocrude oil followed by catalytic upgrading, by novel hydrothermal upgrading (HTU) getting biocrude with low content of oxygen for further upgrading. Currently, because of promoting the use of biofuels, diesel is a mixture of petroleum hydrocarbon fractions, and fatty acid methyl esters (FAME). On the market there are the most common fuel oils with a content of 7 and 20% vol. FAME. The research results has shown that diesel fuel can be formulated using other biocomponents like higher alcohols such as biobutanol, pure vegetable oil. Biofuels should have a high cetane number, high calorific value, normal rheological properties and proper viscosity. Also important is the fuel spray and its evaporation in the engine. Biofuel quality with many of biocomponents should be thoroughly investigated, because the individual components may interact with fuel system materials. For example support of proper emulsifiers are needed in diesel fuel containing ethanol, because it allows to preserve of uniformity of fuel at low temperatures. Synthetic biohydrocarbons are chemical compounds with similar physico-chemical properties characteristic for middle distillates used in the production of petroleum diesel. By changing the conventional fuel formula one have to be aware of the requirements posed by modern fuels car engines, in this respect must be maintained full compliance requirements.

Biofuels considered in this work were compared with diesel oil from the point of view of various physico-chemical limits included in EN 590: 2009 standard and the new technical specification under development by CEN (European Committee for Standardization) under the title: Automotive fuels - Paraffinic diesel fuel from synthesis or hydrotreatment- requirements and test methods, and based on recommendations of the world's car companies members of the Committee for the Affairs of the Worldwide Fuel Charter (WWFC). Figure 9 summarized the technologies used for production biofuels discussed in this chapter. Some of them like HVO diesel (NExBTL) are already commercially available, and others are at demo stage.

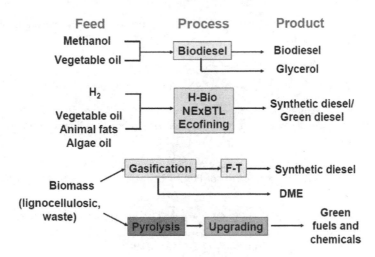

Figure 9. Current (green rectangle) and future (the blue and pink rectangles) most promising ways for producing of second generation biodiesel [19]

Author details

Artur Malinowski, Joanna Czarnocka and Krzysztof Biernat

Automotive Industry Institute, Department for Fuels and Renewable Energy, Warsaw, Poland

References

[1] IEA. International Energy Agency. See also: http://biofuels.abc-energy.at/demoplants

[2] Duncan R. C.: Evolution, technology, and the natural environment: A unified theory of human history. Proceedings of the Annual Meeting, American Society of Engineering Educators: Science, Technology, & Society, 1989, 14B1-11 to 14B1-20.

[3] Directive 2009/28/EC of the European Parliament and of the Council of 23 April 2009 on the promotion of the use of energy from renewable sources and amending and subsequently repealing Directives 2001/77/EC and 2003/30/EC

[4] S.N. Naik , Vaibhav V. Goud, Prasant K. Rout , Ajay K. Dalai, Production of first and second generation biofuels: A comprehensive review, Renewable and Sustainable Energy Reviews 14 (2010) 578–597

[5] Sinha,S., Misra, NC, Diesel fuel alternative from vegetable oil. Chem Eng World 1997; 32 (10):77-80

[6] Laza,T., Bereczky,A., Basic fuel properties of rapeseed oil-higher alcohols blends. Fuel 90 (2011) 803-810

[7] C. Jin et al., Renew. and Sustain. Energy Rev. 15 (2011) 4080-4106

[8] EN 590:2009. Automotive fuels-Diesel-Requirements and test methods

[9] Czarnocka J, Malinowski A, Sitnik L, *Assessment of physicochemical properties ternary bio-fuels to power diesel engines*, The Archives of Automotive Engineering 4/2011 p. 37-42

[10] Sitnik, L., New ecofuel for diesel engines, Journal of POLISH CIMAC, Science publication of Editoral Advisory Board of POLISH CIMAC, ISSN 1231-3998, pp 155-159, Gdansk 2009

[11] Rakopoulos DC, Rakopoulos CD, Giakoumis EG, Dimaratos AM, Kyritsis DC. Effects of butanol–diesel fuel blends on the performance and emissions of a high-speed DI diesel engine. Energy Convers Manage 2010;51: 1989–97.

[12] Fundamentals of Thermochemical Biomass Conversion. eds. R. P. Overend, T. A. Milne, and L. K. Mudge. Elsevier Applied Science, New York, NY, 1985

[13] Zhang, Y., Riskowski, G., Funk, T., "Thermochemical Conversion of Swine Manure to Produce Fuel and Reduce Waste," Technical Report to Illinois Council on Food and Agricultural Research , University of Illinois at Urbana-Champaign, 1999

[14] G. Centi, P. Lanzafame, S. Perathoner, Analysis of the alternative routes in the catalytic transformation of lignocellulosic materials, Catalysis Today 167 (2011) 14–30

[15] J.A. Petri, T.L. Marker, US Patent 7,511,181 (2009).

[16] J. Holmgren, C. Gosling, R. Marinangeli, T. Marker, Hydrocarbon Process. 9 (2007) 67.

[17] E. Koivusalmi, J. Jakkula, US Patent 7,459,597 B2 (2008).

[18] A.R. Pinho, M. Silva, A.P. Silva Neto, J.A.R. Cabral, US Patent, US 7540952 B2 (2009).

[19] T. Kalnes et. al., 1st Alternative Fuels Technology Conference, February 18, 2008, Prague, Czech Republic

Chapter 7

Simulation of Biofuels Combustion in Diesel Engines

Andrey Marchenko, Alexandr Osetrov,
Oleg Linkov and Dmitry Samoilenko

Additional information is available at the end of the chapter

1. Introduction

In the study of the working process, the development of new engine construction or modernization of an existing one is necessary to use simulation with mathematical models. Modeling of the processes inside the cylinder allows in a first approximation to evaluate engine performance, choose the rational value of adjustment or constructive parameter, to reduce material, labor and time required to conduct experimental research.

One of the most difficult process for simulation is the combustion process in diesel engines. This process is determined and accompanied by a number of other processes and phenomena. There is intense interaction between the motion of the fuel jets and air flow in the cylinder, heat transfer between the combustion chamber zones and walls, volume evaporation from the surface of liquid droplets. All this leads to the formation of the active nucleus of the fuel oxidation and its ignition, volumetric and then the diffusion combustion.

Currently, there are a number of hypotheses about the behavior of each of these processes and their interaction. For each hypothesis proposed mathematical description of a different degree of accuracy.

The most complex model implemented technology of Computational Fluid Dynamic (CFD) - three-dimensional simulation of gas flow and the injected fuel in the cylinders and manifolds of internal combustion engines [1-5]. The most popular programs are: KIVA (Los Alamos National Laboratory, Los Alamos, New Mexico); STAR-CD (CD-adapco, headquarter Melville, New York, USA); FIRE (AVL, headquarter in Gratz, Austria); VECTIS (Ricardo, headquarter Shoreham-by-Sea, England, United Kingdom).

For example, the software package AVL FIRE Engine includes over 20 different models of formation and spread of the jet, its decay, crushing drops, collisions between them, the evaporation of fuel and its interaction with the wall of the combustion chamber [6]. The formation of liquid films, their distribution and evaporation, the interaction with the walls and the liquid fuel torches are also simulated. Several models describe the processes of ignition, combustion and the formation of harmful substances, taking into account detailed chemical kinetics of reacting systems.

A significant technical challenge of CFD models is the complexity of calculations and the need for powerful computers. Data preparation only for one simulation with highly skilled personal could takes a few days. Calculation time for one variant of the engine - a few hours and sometimes days. Implementation of these programs for optimization calculations is problematic because optimization process has to count thousands of design options.

Thermodynamic and phenomenological models that use the 0 - or 1-dimensional representations, require less time and resources. The most popular programs were GT-Power (Gamma Technologies, Inc, headquarter Westmont, Illinois, USA), BOOST (AVL,Gratz, Austria), WAVE (Ricardo, Shoreham-by-Sea, England, United Kingdom), DIESEL RK (Moscow State Technical University named after Bauman, Moscow, Russian Federation). These software products usually include a one-dimensional model of gas exchange. To calculate the mixing and combustion in a diesel engine used empirical or semi-empirical models [7-11].

The most sophisticated models of combustion used in thermodynamic models are models of H. Hiroyasu [9], as well as Razleytsev N.F. and Kuleshov A.S. models[7, 8]. In these models, the propagation of fuel jet is described by the criterial equations obtained on the basis of experimental data. It has been assumed in this modelsthat the main influence on the rate of heat generation rate has drops evaporation rate and the speed of the air penetrated in the combustion zone. Also, the effects of air swirl on the development of fuel sprays is considered. In models of mixing, combustion and evaporation using an average diameter of the droplet on the Sauter. The fuel jet is considered as a set of zones, each of which has a characteristic temperature, the volume, fuel-to-air ratio.

These models allow us to investigate the influence on the combustion of compression, timing and duration of the injection, hole diameter and the number of sprays in the fuel injector, characteristics of fuel injection, combustion chamber shape, correlate the direction of fuel jets with combustion chamber and swirl intensity, take into account the interaction of jet fuel with the walls and to each other and finally allow you to perform multi-factor multi-criteria optimization.

However, the use of this class of models requires detailed design information of the simulated engine, setting up empirical relations and coefficients to make a relatively labor-intensive verification.

Widespread empirical or semi-empirical models of combustion, which describe the geometric shape of the heat generation curve [10-15] (second group) are also presented. Such models are easy to describe and versatility of use. For example, in a model of prof. VibeI.I. [10], the rate of combustion and the proportion of burnt fuel are described by semi-empirical dependencies:

$$\frac{dx}{d\varphi} = -C\frac{m+1}{\varphi_z}\varphi^m \exp(C\overline{\varphi}^{-m+1}); \tag{1}$$

$$x = 1 - \exp(C\overline{\varphi}^{-m+1}), \tag{2}$$

where $\overline{\varphi} = \varphi/\varphi_z$, φ, φ_z - respectively, relative duration of combustion, the current duration of combustion from the start of combustion and combustion duration shown in angles of rotation of the crankshaft;

C - constant (for example, at the end of the combustion when $x = x_z = 0.999$, $C = \ln (1-0,999) = -6,908$);

m-index of combustion character.

Feature of empirical models is that all input values of the calculation formulas are constant values and are given by experimental data or chosen from the recommended by investigators ranges. For example, in first approximation, prof. Vibe I.I. recommends$0 \le m \le 0,7$for diesel engines, and in the work of scientists from Bauman Moscow state University(Moscow, Russian Federation) values of m range from -0.3 to 0.7.

Use of this class of models suitable for describing the combustion in a specific engine running on one mode of his work.When changing a constructive parameter and adjusting the engine or the conditions of his work empirical models stop producing an accurate result.

The drawback of empirical models of combustion is the complexity of their use in calculations of the harmful substances formation in diesel engines, in particular nitrogen oxides. NO output in accordance with the thermal theory of ZeldovichU.B. [16] is extremely sensitive to the magnitude of the temperature in the cylinder. Therefore, in these calculations, it is important to accurately determine the temperature and, consequently, the heat generation curve. This curve, calculated by the empirical models as a rule have one peak that does not comply with the combustion process in diesel engines for most modes of operation. Accordingly, the accuracy of the calculation output of harmful substances by using models of this class is relatively low.

Most of the problems that arise in the practice of design and research of various diesel engines can be solved using "intermediate" type models [11, 18-20] (third group). These models combine the advantages of computational methods from first and second groups.

A number of models describes the combustion process by using Vibe I.I. relationships (1) and (2) [19, 20], but unlike empirical models the indices of combustion duration φ_z and combustion character m are functions of design parameters and operation modes.

The data obtained by processing the experimental indicator diagrams, confirm the correctness of this approach (Fig. 1)

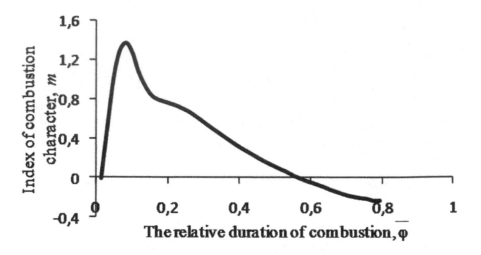

Figure 1. The change in the index of combustion character *m* during operation cycle in the 4 stroke autotractor diesel engine with turbocharger (SMD-23)

According to the variable nature of the index of combustion character*m* for the differentiation of equation (2) the next dependence was obtained which is different from equation (1):

$$\frac{dx}{d\varphi} = -C\exp(C\bar{\varphi}^{m+1})\frac{1}{\varphi_z}\left[(m+1)\bar{\varphi}^m + \bar{\varphi}^{m+1}\ln\bar{\varphi}\frac{dm}{d\bar{\varphi}}\right]$$

Filipkovsky A. I.proposed to determine the index of combustion character*m* and the duration of the combustion $\varphi_z$in Vibe I.I. dependencies (1) and (2) as a function of the parameters of the evaporation, diffusion and chemical kinetics of reaction [19]. The model takes into account the main factors that determine the combustion process:

• design features of the combustion chamber (chamber shape, the diameter of the cylinder and the neck chamber, swirl ratio);

• characteristics of the fuel injection and atomization (diameter and the effective cross section of nozzle holes, duration, and mean pressure of injection, amount the fuel during operation cycle, the physical characteristics of the fuel);

• thermo-and gas-dynamic parameters of the charge in cylinder (pressure and density of charge at the end of a conditional extended to top dead centre (TDC) compression, the tangential velocity of the charge in the combustion chamber);

• mode parameters of the engine (speed, excess air ratio).

The model assumed that the development of chain reactions begins with the start of fuel injection into the diesel cylinder, rather than the beginning of combustion, as in the model of prof. VibeI. I. The curve of heat generation rate, calculated by the model, has one peak.

Calculations of heat curves by the model of Filipkovsky A.I. for medium-speed four-stroke diesel engines with turbocharging, 26 cm bore and stroke 34 cmand four-stroke diesel engines with turbocharging, 32 cm bore and stroke 32 cm with volume mixing processes have shown good agreement with experimental data. However, practical application of this model for high-speed automotive diesel with a volume-film-mixing processes did not produce positive results. The discrepancy between the calculated and experimental data is greatest in the partial modes, where the curve has a two-peak heat generation rate in nature (Fig. 2). Finally, this method gives a significant error in the calculations for biofuels because of significant differences in the physicochemical properties of diesel fuels and biofuels.

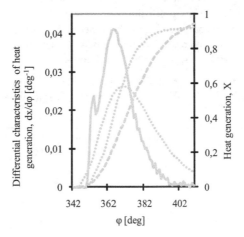

Figure 2. Comparison of experimental and calculated heat generation characteristics of the ethyl esters of rapeseed oil ⋯⋯⋯⋯⋯– Calculation by the method of Dr. Filippovsky A.I.; ————– Experimental data

Despite these problems, the Filipkovsky A.I. model, in our opinion, has the potential for further improvement. Obviously, it is necessary to adapt this model to integrate features of medium-speed diesel engines, physicochemical properties of biofuels, as well as operation modes of small and medium loads, where the heat generation rate has two-peak character.

Significant influence on the combustion process have physical and chemical properties of fuel. In present study, the features of the processes in the engine cylinder associated with the use of bio-fuels of plant-based origin, in particular mixtures of rapeseed oil (RO) with diesel fuel (DF) and the ethyl ester of rapeseed oil (EERO). In conducted by authors experimental studies have shown that the presence of oxygen in the molecule of biofuels will intensify the process of diffusion combustion, which should be considered when developing a mathematical model.

This chapter describes the results of experimental studies of biofuels in diesel engines, the mathematical model of combustion in the diesel enginecylinder and the results of verification.

2. Experimental studies of biofuels in diesel engines

Experimental studies are needed to obtain basic data for modeling, getting a number of empirical coefficients in the model equations and refinement of physical laws, comparison of experimental and calculated data.

2.1. Investigation of physicochemical properties of biofuels

Physicochemical properties of the investigated biofuels are presented in Table 1.

Analysis of the data in Table. 1 shows that the properties of plant-based fuels are significantly different from the properties of diesel fuel: PM and EEROafter comparison with DFhave respectively 14 and 13.5% less low heat values, for 10 and 8.1% higher density, for 14.1 and 21.9% higher surface tension, and for 22.8 and 8.5 times higher viscosity. For the combustion of 1 kg of RO and EERO required respectively 12.7 and 12.6% less air, which is associated with the presence of oxygen in the structures of their molecules.

It should be noted that the trial set of EERO contained unreacted raped oil, so the physical and chemical properties of ethyl differ are different from those given in the technical literature [21].

Property	Diesel fuel (DF)	Rape oil (RO)	Ethyl ester of rapeseed oil (EERO)	Mixtures		
				DF: RO (3:1)	DF: RO (1:1)	DF: RO (1:3)
Elemental composition, %:						
carbon (C)	87	77,9	77,6	84,5	82	79,8
hydrogen (H)	12,6	11,9	12	12,3	12	11,8
oxygen (O)	0,4	10,2	10,4	3,2	5,9	8,5
sulfur (S)	0,04	0	0	0,03	0,02	0,01
The amount of air for combustion per mass unit of fuel Lo, kg / kg	14,4	12,7	12,7	13,9	13,5	12,9
High heat value Qv, MJ / kg	44,95	39,3	39,2	43,4	42,0	40,6
Low heat value of Qn, MJ / kg	42,2	36,8	36,9	40,7	39,3	38,0
The density ρ, g/m3 (20 ° C)	825	915	895	849	872	894
The kinematic viscosity v, mm 3 / s (20 ° C)	3,8	87	32,48	26,3	47,6	67,8
The surface tension σ * 103 N / m (20 ° C)	28,9	33,3	36	30,1	31,2	32,3

Table 1. Physicochemical properties of biofuels.

The difference between the physical and chemical properties of biofuels on the properties of diesel fuel is the cause of changes in diesel working process and performance, which should be considered when simulating processes inside the cylinder.

2.2. Studies of dispersion atomized biofuels

To clarify the empirical and criterial relationships that characterize the quality of atomization of biofuels, an experimental study was made of atomizationdispersion.

Single injections were made on glass plates coated with a layer of soot and kerosene and on a top side covered with a layer of magnesium oxide (which has a bright white color) for clarity of prints fuel droplets.

The studies were conducted on the following frequencies of high pressure fuel pump camshaft rotation: 900,700 and 770 rpm. Fuel rack setting was made for maximum fuel delivery.

Micrographswere obtained in the experimental study of dispersion of the atomization of various fuels and shown on Fig. 3.

Photomicrographs are processed according to the procedure [22]. The relative fuel atomization characteristics were obtained as the results (Figures 4-7): Differential (R0 - quantitative; R2 - surface; R3 - volume) and integral (S0 - quantitative; S2 - surface; S3 - volume).

The data obtained allowed to estimate the average diameter of fuel droplets of different composition.

The most commonly used parameter for calculating the evaporation of fuel is the average volume-surface droplet diameter (Sauter diameter):

$$d_{32} = E_{32} d_c M^{0.0733} / \left(\rho We \right)^{0.266} , \tag{3}$$

where E_{32} is a constant factor depending on the design of the nozzle and the method of averaging the droplet size;

d_c-diameter of atomized holes;

M-criterion, which characterizes the ratio of surface tension and viscosity;

W_e- Weber criterion;

ρ- air density to fuel ratrio

Calculations of Sauter diameter of droplets for a four-stroke auto-tractor turbocharged diesel engine, which has a cylinder diameter 120 mm and 140 mm stroke running on standard diesel fuel using the standard fuel system show that the value of d_{32}on nominal power mode ranges from 26 to 29 microns. Greater droplet diameter values obtained in experimental studies (Fig. 4), due to the fact that the injection was made into the environment under atmospheric conditions. It is obvious that in a running diesel engine a high temperature of the charge in the cylinder causes a greater atomization and evaporation of fuel droplets.

Therefore, when refinement dependencies (3) for the case of biofuels, relative (not absolute) values of d_{32}(Table 2) was used. Analysis of the data in Table. 2 shows that the dependence (3) with appreciable error describes the variation of d_{32} for plant-based fuels. Authors have proposed an empirical correction for the dependence (3), depending on the viscosity of the fuel and allows to do more accurate calculation of the average volume-surface diameter of drops:

$$k_f(v) = -0,00010939 \cdot v^2 + 0,0052066 \cdot v + 0,98179447. \qquad (4)$$

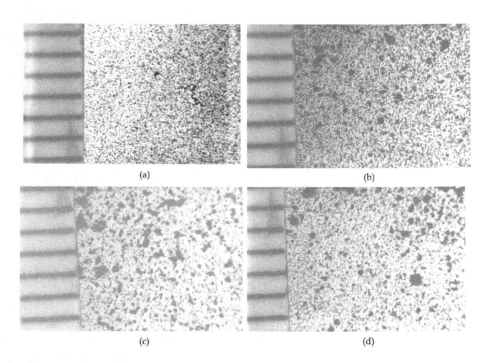

(a)   (b)

(c)   (d)

Figure 3. Micrograph of atomized fuel droplets (n = 900 rpm): a standard diesel fuel (a) a mixture of diesel fuel and rapeseed oil in the ratio 1:1 (b); pure rapeseed oil (c); ethyl ester of rapeseed oil (d)

	Diesel Fuel (DF)	Mixture DF: RO (1:1)	Rape Oil (RO)	Ethyl ester of rapeseed oil(EERO)
$\bar{d}_{32}$(experiment)	1,00	1,509	2,000	1,877
$\bar{d}_{32}$(calculated by the formula (3))	1,00	1,537	3,176	1,488
$\bar{d}_{32}$(adjusted value)	1,00	1,509	1,927	1,541

Table 2. The relative diameters of the droplets of different fuels

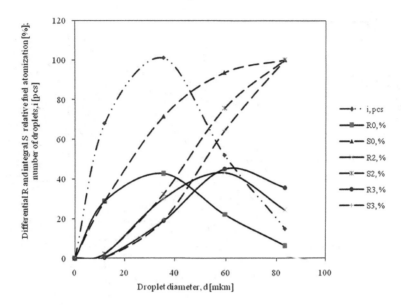

Figure 4. The relative atomization characteristics of diesel fuel (high pressure fuel pump camshaft rotation speed 900 rpm)

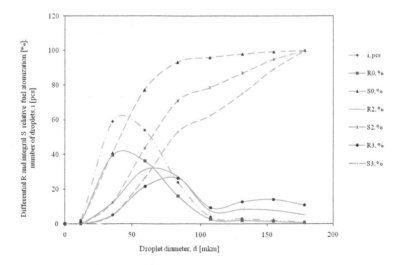

Figure 5. The relative characteristics of spray mixture DF and RO (1:1) (high pressure fuel pump camshaft rotation speed 900 rpm)

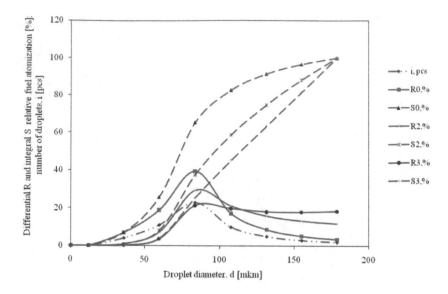

Figure 6. The relative atomization characteristics of RO (high pressure fuel pump camshaft rotation speed 900 rpm)

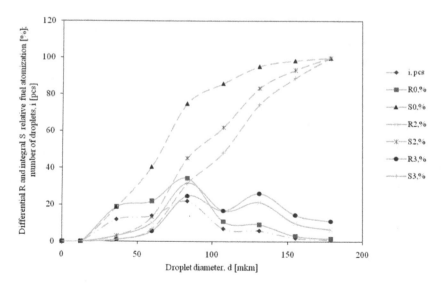

Figure 7. The relative atomization characteristics EERO (high pressure fuel pump camshaft rotation speed 900 rpm)

Dependence (3), corrected to (4) becomes:

$$d_{32} = \frac{k_f \cdot d_c \cdot E_{32} \cdot M^{0.0733}}{\left(\rho \cdot We\right)^{0.266}}. \tag{5}$$

2.3. Experimental investigations of biofuels implementation in diesel

Experimental studies of the engine running on traditional and biofuels performed on a test bench with a diesel engine SMD-23, equipped with a turbocharged and intercooler system. Brief technical characteristics of a diesel engine SMD-23 is shown in Table. 3, the picture of experimental facility is shown in Fig. 8.

Parameter	Value
Number of cylinders	4
Bore, mm	120
Stroke, mm	140
The geometric compression ratio	15,5
Rated power, kW	120
Rated speed rpm	2000

Table 3. Summary of technical characteristics of a diesel engine SMD-23

Figure 8. Experimental stand

Engine tests were conducted on the modes of engine load characteristic related to rated power mode with engine speed 2000rpm and peak torque mode with engine speed 1500 rpm.

During the tests on each mode, the parameters of air and fuel delivery systems, exhaust gas, coolant and oil were measured. Engine speed and torque were also detected. Indexing, the definition of stroke in injector idle and measuring the pressure in the fuel injection pipe was carried out. Also emissions were measured and included NO_2, CO and smoke registration.

The values of injection timing angle and adjust fuel pump adjustments remained unchanged.

Main diesel indices that running on a different fuels are shown in Fig. 9.Lets consider the effect of physicochemical properties of bio-fuels on the performance of diesel.

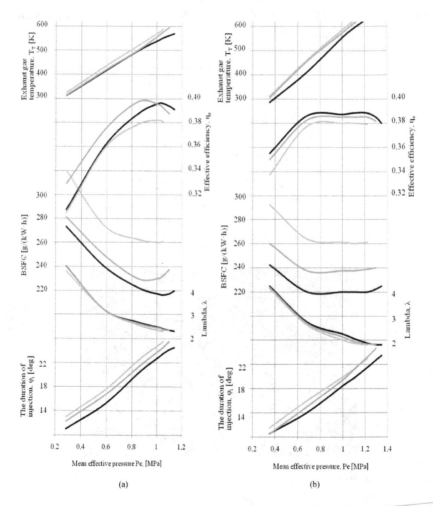

Figure 9. Effect of load on the performance of diesel exhaust gas SMD-23: engine speed 2000 rpm (a); engine speed 1500 rpm (b) ▬ -DF; ▬▬ - EERO; ▬▬RO: DF (1:1)

Injected fuel and air mixing. Injection of plant-based fuel into the combustion chamber is carried out with the higher maximum pressure than the injection of diesel fuel, which is explained by the influence of lower compressibility and higher viscosity of plant-based fuels.

Greater surface tension force and greater kinematic viscosity (see Table. 1) provide more later decay of injected plant-based fuel on the droplets and formation of smaller atomizing cone in comparison with diesel fuel. This dramatically increases the diameter of fuel droplets. As shown in [21, 23, 24], atomizing cone is reduced by 10% (using mthyl ester of rapeseed oil) and the average volume-surface droplet diameter d32 increases, respectively, in 1,5-1,877 times (when using pure RO, a mixture of RO and DF (1 to 1) and EERO).

Injection of plant-based fuels with the higher maximum pressures greater diameters of droplets in combination with a greater specific weight increase penetrating power and the range of fuel jet. The duration of injection of plant-based fuels φ_i increases slightly (1-2 crank angle) as a result of significant increase in injection pressure with a small increase in a fuel delivery.

The above factors lead to the fact that in the case of using plant-based fuels the volume fraction of mixing is reduced and the fraction of wall-film mixing is increase. The quality of volume mixing become lowered in this case.

The period of ignition delay. As a result of processing the experimental indicator diagrams, integral and differential characteristics of heat generation in the cylinder were obtained over the entire range of investigated fuel mixtures and regime characteristics of the engine. In Fig. 10 shows the relative heat generation characteristics when burned pure diesel fuel, and mixtures EERO,RO: DF (1:1) on high load (Pe = 1.25 MPa and Pe = 1.01 MPa) and medium (Pe = 0.57 MPa and 0 67 MPa) load at engine speed 1500 rpm and engine speed 2000 rpm. From the analysis of these characteristics is difficult to see any legitimate differences in the ignition delay period of plant-based fuels and diesel fuel.

Consequently, we can conclude that the flammability of plant-based fuels is almost unchanged in comparison with diesel fuel flammability.

The first peak of heat generation rate. As we can see from Fig. 10, on the most modes the maximum heat generation rate for plant-based fuel in this period is lower than for diesel fuel. In addition, the area under the first peak of the curve dx / dφ smaller, and hence smaller the amount of fuel burned out in this period.

This fact is obviously related to the deterioration of the mixing between the ignition delay when using mixtures of RO with the DF and EERO. Reduction in the angle of frame divergence, increasing the relative amount of fuel that enters the wall of the combustion chamber, a significant increase in the average diameter of droplets leads to a deterioration in mixing formation and reduce the relative amount of fuel vaporized during the period of ignition delay.

The second peak of heat generation rate. After burning the fuel, evaporated during the period of ignition delay, there is a diffusion combustion of the fuel droplets in the fuel torch, as well as the fuel evaporating from the walls of the combustion chamber after the contact of the torch and the wall. The nature of the combustion process in this period determines the indicator performance of the cycle [25].

In an experimental engine used a cylindrical combustion chamber, that implements volume-film mixture formation. Obtained data is contradictory ex facte (see Fig. 10). Increasing the amount of fuel reaching the wall, large diameter drops, the heterogeneity of atomization when using plant-based fuels should lead to a decrease in the rate of evaporation and combustion of fuel, especially in the modes of small loads, when the wall has a lower temperature. However, it is clear that almost on all modes there is an increase in the rate of combustion as compared to DF. The deterioration in the amount of mixing in this case does not lead to a decrease in combustion rate and it can be seen not only in the modes of high loads, but also in modes of low loads.

Intensification of the diffusion combustion of plant-basedfuels, can obviously be explained by the presence of oxygen in the structure of the molecule. When burning fuel droplets of biofuel, the oxygen is in the molecule of fuel. This oxygen is more active than molecular oxygen. That is why, even at low temperatures of plant-based fuel oxidation rate of its "own" oxygen is very high. All this probably leads to an increase in diffusion combustion rate in general.

The increase in the rate of combustion of plant-based fuels in the main period of combustion in most cases leads to a slight increase in average temperatures and pressures in the cylinder. In addition, the exhaust gas temperature rise in the exhaust manifold (see Fig. 9.).

The period of slow combustion. During this period there was burning of fuel in the cylinder. From Fig. 10 difficult to see the end of the combustion of different fuels. However, it is clear that the differences between the test fuel at the end of the combustion is low. Accelerated burning of plant-based fuels, during the second period of combustion, apparently compensate by slow combustion in the first period. So the total duration of combustion is practically unchanged.

Effective performance. As it can be seen from Fig. 9, the use of plant-based fuels leads to an increase in break specific fuel consumption because of reduction in their low heat value compared to diesel fuel. Changing in effective efficiency of the diesel engine is not so clear.

The increase in the rate of diffusion combustion, high quality film mixing on the high loads modes lead to an increase in the effective efficiency if we use plant-based fuels.

At low load modes mixture formation deteriorates in the volume of the combustion chamber. In addition, the share of plant – based fuel burning in the relatively cold wall surface areas of combustion chamber, which explains the decrease in the effective efficiency at low load modes.

The toxicity of exhaust gases. When using supplements of vegetable-based oils to diesel fuel, and if diesel engine operating on pure EERO on modes of high and medium loads the smokeemission reduced in 1.6-2 times and NO emissions increase for 5-15%. In most modes of low-load smoke and NO are reduced (in 1.2-2 times), or remain unchanged. CO emissions using different fuels are comparable.

Experimental studies have provided initial information on the physicochemical properties of plant-based fuels, low of flow injection, atomization, mixing and combustion in the cylinder, the data for mathematical modeling processesinside the cylinder.

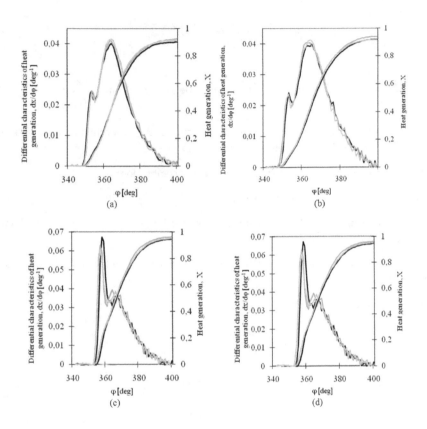

Figure 10. Heat generation characteristics in the diesel cylinder: engine speed 1500 rpm: Pe = 1.25 MPa (a) Pe = 0.67 MPa (b); engine speed 2000 rpm: *Pe* = 1,01 MPa (c), *Pe* = 0,57 MPa (d); ▬▬ -DF; ▬▬ - EERO; ▬▬RO: DF (1:1)

These differences in the physicochemical properties of biofuels and their impact on flow processes in the cylinder of a diesel engine form the basis for the developed mathematical model of combustion.

3. Description of the proposed mathematical model of combustion

Lets consider the features of the proposed model of combustion.

Differential characteristic of heat generation proposed to describe by two curves corresponding to the periods of ignition (or "fast" combustion) and the diffusion combustion

$$\left(\frac{dx}{d\varphi}\right)_I = -A \cdot C \cdot \exp(C \cdot \overline{\varphi}_I^{-m_I+1}) \frac{6n}{\varphi_{zI}} \left[(m_I+1)\overline{\varphi}_I^{-m_I} + \overline{\varphi}_I^{-m_I+1} \ln \overline{\varphi}_I \frac{\overline{dm_I}}{d\varphi_I} \right] \tag{6}$$

$$\left(\frac{dx}{d\varphi}\right)_{II} = -C \cdot \xi_v \cdot S \cdot \exp(C\overline{\varphi}_{II}^{-m_{II}+1}) \frac{6n}{\varphi_{z_{II}}} \left[(m_{II}+1)\overline{\varphi}_{II}^{-m_{II}} + \overline{\varphi}_{II}^{-m_{II}+1} \ln \overline{\varphi}_{II} \frac{\overline{dm_{II}}}{d\varphi_{II}} \right] \tag{7}$$

where A - coefficient taking into account the influence of the proportion of vaporized fuel during the ignition delay at the rate of fast combustion;

C - coefficient taking into account the completeness of combustion;

ξv - the degree of efficient use of air charge;

S - coefficient taking into account the share of fuel burned for the period of the ignition (linking the two periods).

The index «I» related to parameters that identified ignition, the index «II» - the process of diffusion combustion.

Dynamics indicators for the respective periods of combustion:

$$m_I = 4 \cdot \overline{\varphi}_{mI} \cdot \left(1 - \overline{\varphi}^{-\overline{\varphi}_{mI}} \right); \tag{8}$$

$$m_{II} = 9 \cdot \overline{\varphi}_{mII} \cdot \left(1 - \overline{\varphi}^{-\overline{\varphi}_{mII}} \right), \tag{9}$$

where $-\overline{\varphi}_{mI}$ and $\overline{\varphi}_{mII}$ relative moments of maximum heat generate rate;

$\overline{\varphi}_I$ and $\overline{\varphi}_{II}$ - the relative angles of the crankshaftrotation: $\overline{\varphi}_I = \varphi / \varphi_{ZI}$, $\overline{\varphi}_{II} = \varphi / \varphi_{ZII}$;

φ - the current angle of the crankshaft rotation from the start of combustion;

φ_{ZI}, φ_{ZII} - respectively, the duration of fast and diffusive combustion.

At each calculated section the values $\left(\frac{dx}{d\varphi}\right)_I$ and $\left(\frac{dx}{d\varphi}\right)_{II}$ was comparing. The final calculated heat generation rate $\frac{dx}{d\varphi}$ took the value of greater meaning of two rates.

The total amount of burnt fuel are determined by integrating the function dx / dφ in the area of combustion

$$x = \int_{\varphi_N}^{\varphi_k} \frac{dx}{d\varphi} d\varphi, \tag{10}$$

where φ_N, φ_k - respectively the beginning and the end of combustion.

In formulas (6) - (9) there are the parameters A and, φ_{ZI} and φ_{ZII}, ξ_v, which, unlike the parameters of the known formulas Vibe I.I. accounted specific processes of fuel injection, mixing, evaporation, combustion, and the interaction of these processes with each other.

In generalizing the data obtained by processing the experimental indicator diagrams, empirical correlations was proposed to determine the relative moment of maximum heat generation rate during periods of combustion:

$$\overline{\varphi}_{mI} = 0,8 + \frac{0.03 \cdot b_e \cdot \varphi_{ZI}}{6 \cdot n}; \overline{\varphi}_{mII} = 0,16 + \frac{0.03 \cdot b_e \cdot \varphi_{ZII}}{6 \cdot n}, \tag{11}$$

where b_u- a constant of relative evaporation.

The relative constant of evaporation

$$b_u = K_u / d_{32}^2 \tag{12}$$

where K_u - constant of evaporation, calculated for the average diameter d_{32} of the droplet under Sauter.

Prof. Razleytsev [8] estimated, that during the evaporation of fuel in a diesel engine cylinder the average evaporation constant:

$$K_{uT} = (10^6 p_c)^{-1} \tag{13}$$

where p_c - the pressure in the cylinder at the end of a conditional extension to TDC compression.

Theoretical constant K_{uT} does not include an increase in the rate of evaporation of droplets during combustion, the effect of size of drops, speed and frequency of turbulent vortices arising in the diesel cylinder. This dependence is in practical calculations can be taken into account by correction function Υ:

$$Ku = \Upsilon \cdot K_{uT}. \tag{14}$$

In [8] proposed the following formula for determining the correction function:

$$Y = y \left(W_T d_{32} \right)^{0.75} p_c^{\ 0.25}, \tag{15}$$

where y - constant empirical coefficient depending on the design of the combustion chamber and taking into account the effect of unaccounted secondary factors.

W_T - the tangential velocity of the charge in the combustion chamber;

p_c - the calculated pressure at the end of a conditional extension to TDC compression.

It should pay particular attention to the coefficient of y. It is obvious that there are permanent factors that defined y coefficient - engine design, adjustments and settings mode. On the other hand, while using different fuels - the value of this ratio will be determined by physical and chemical properties of fuels.

As shown in Section 2, an important property of plant-based fuelsthat have a material effect on the combustion process is the oxygen content in the molecule. Increasing the number of bound oxygen in the molecule leads to an increase in the rate of diffusion combustion. Accordingly, the simulation of combustion is expedient to increase the coefficient y proportional to the share of the oxygen in the molecule of fuel.

In this study, a constant value y was adjusted for each fuel type on the basis of providing the best agreement between calculated and experimental data. For all the calculations for one type of fuel y constant has not changed.

In accordance with the original model [19] the duration of diffusion combustion:

$$\tau_{zII} = \varphi_i + \varphi_b, \tag{16}$$

where φ_i - the duration of fuel injection;

φ_b - the duration of burn-out fuel after the injection.

The duration of burn-out fuel ϕ_b characterized by the time of evaporation and combustion of large droplets delivered in the diesel engine cylinder at the end of injection. This time depends on the fineness of atomization, the distribution of drops, the parameters of the working fluid in the cylinder, air-fuel ratio, etc. ϕ_b can be calculatedfrom the formula [8]:

$$\varphi_b = K_\alpha \cdot \varphi_e, \tag{17}$$

where ϕ_e - the duration of the evaporation of large droplets of fuel;

K_α - correction function which takes into account the time of fuel vapors burning.

The duration the large droplets evaporation of fuel:

$$\varphi_e = \frac{d_K^2}{K_u} \tag{18}$$

where d_K - an average diameter of large drops of fuel injected into cylinder by the end of the fuel delivery.

In [8] proposed to determine the diameter of the large drops by the formula:

$$d_K = B \cdot d_{32} \tag{19}$$

In this formula, the size factor is:

$$B = 1.5 + 0.018 \exp\left(\Delta p_{fi}^{0.272}\right), \tag{20}$$

where Δp_{fi}- the average pressure drop during injection, MPa.

Correction function at the time of burning-out of fuel vapors can be determined from the dependence [8] :

$$K_\lambda = 1 + \frac{A_3 K_u}{(\lambda - 1)} \tag{21}$$

where λ- is the ratio of actual air-to-fuel ratio to stoichiometry for a given mixture;

A_3 - coefficient, which is determined by identifying a number of experimental data for defined row of engines and can be taken equal to $2.5 \bullet 10^6$.

We proposed to determine the duration of the fast combustion as a function of the duration of the ignition delay period:

$$\tau_{zI} = \tau_i \cdot K_\lambda, \tag{22}$$

where τ_i - the period of ignition delay in seconds.

If to go to the crank angle, the duration of fast and diffusive combustion are determined by the following formulas:

$$\varphi_{ZI} = \tau_{ZI} 6n; \varphi_{ZII} = \theta_i + \varphi_b 6n, \tag{23}$$

where θ_i - injection advanced angle.

In developed model has been assumed that burn rate during the ignition mostly depends on the amount of fuel vaporized during the period of ignition delay. In turn, the calculation of the first peak heat generation rate coefficient taking into account the influence of the proportion of vaporized fuel during the ignition delay period.

$$A = K_I \cdot \sigma_I \tag{24}$$

where σ_I - the relative amount of fuel injected during ignition delay period φ_I;

K_I - coefficient of proportionality.

In [23] proposed the dynamics of heat generation during diffusion combustion (equation (7)) to adjust with the ratio ξv, which is a degree of efficient use of air charge in cylinder:

$$\xi_V = \frac{\lambda_\partial}{\lambda}, \tag{25}$$

where λ_m - the average λ coefficient in the combustion zone;

λ – the estimated value of λ in the cylinder for full combustion of the fuel injected into the cylinder.

The coefficient ξ_V takes into account the interaction of fuel torch with the wall of the combustion chamber and other factors that reduce the amount of oxidant entering the combustion zone. In [23, 28] describes a method of determining this ratio.

The mathematical model of combustion is integrated into the thermodynamic model of the closed-loop workflow engine with a turbocharger.

4. Implementation of mathematical models for practical calculations

Comparison of calculated and experimental characteristics of heat generation and indicator diagrams using different fuels is shown in Fig. 11-13. It can be seen that the proposed mathematical model provides a satisfactory agreement between the calculated and experimental data in a wide range of biofuels, loads and engine speeds.

A precise description of the combustion process is important in modeling the formation of harmful substances in the cylinder. For example, the error in determining the temperature in the cylinder 80-90K leads to a change in the calculated NO output by 30%, error in determining the temperature of 190 K change in the calculated NO output is 2.7 times [26]. Obviously, using the proposed mathematical model rather than empirical or semi empirical models provide a more accurate calculation of the formation of harmful substances in the diesel engine cylinder.

The adequacy of the developed mathematical model was tested also for its response to the changing influence of parameters - the compression ratio, injection duration and injection delay angle (Fig. 14). It is seen that the obtained numerical data trends and logical and do not conflict with similar data of other researchers [26-28].

We can conclude that the developed mathematical model allows us not only to describe the dynamics of heatgeneration with sufficient accuracy, but also to adequately respond to changes in design and adjustments in the parameters of diesel.

5. Conclusion

Actual and perspective task for modern engine – building has been introduced and solved in the chapter. This task included the development of mathematical model of alternative (biofuels) and fossil fuel (diesel) combustion calculation in the cylinder of diesel engine. It is shown that the physical-chemical properties of biofuels differ significantly from the properties of diesel fuel, which leads to changes in the processes of fuel injection, mixture formation and combustion. All this has a significant impact on the efficient and environmental performance of diesel engines.

Authors have proposed mathematical model that adequately describes the process of combustion of conventional diesel and bio-fuel in the cylinder of diesel engine. It was confirmed by the results of calculation and experimental studies.

The mathematical model proposed by the authors can be used to solve optimization tasks in internal combustion engines running on a diesel fuel, as in this model combustion processes are linked with parameters of engine design and engine working process parameters. The model is developed in a parametric form and reflects the change in design and adjustment parameters of the diesel engine.

An important characteristic of a new mathematical model is an adequate description of the first phase of the combustion process (first peak), which is associated with fuel burn-out, accumulated during the ignition delay, which allows more reliable to calculate the temperature of the working fluid in the cylinder of diesel engine and, consequently, with greater reliability to calculate by Zeldovich methoda number of nitrogen oxides that are formed in the cylinder of diesel engines.

The proposed model can be used in university training programs for professionals in the field of internal combustion engines, as well as in practice of firms participating in the modernization of existing and development of advanced diesel engines.

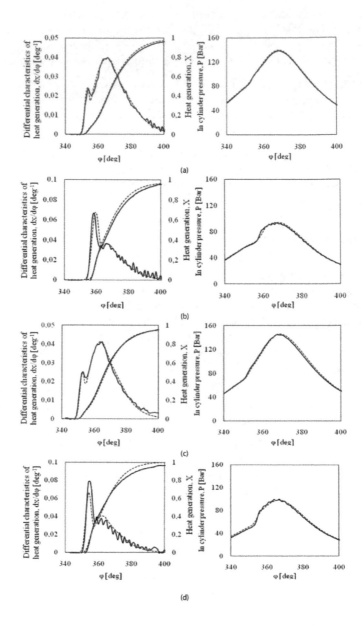

Figure 11. Verification of the model of calculation of heat generation process. Diesel: engine speed 2000 rpm,Pe = 1.1 MPa (a); engine speed 2000 rpm,Pe=0.56 MPa (b); engine speed 1500 rpm, Pe=1.35 MPa (c); engine speed 1500 rpm,pe=0.67 MPa (d); ——————— Experiment; — — — — — — Calculation of the refined model

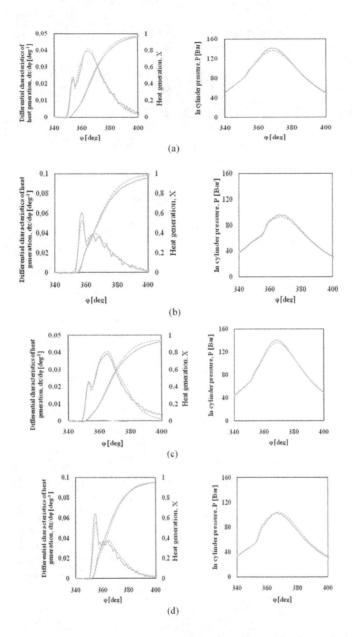

Figure 12. Verification of the model of calculation of heat generation process. A mixture of RO: DF (1:1): engine speed 2000 rpm, p_e =1.1 MPa (a); engine speed 2000 rpm, p_e =0.56 MPa (b); engine speed 1500 rpm, p_e =1.35 MPa (c); engine speed 1500 rpm, p_e =0.67 MPa (d); —————— Experiment; — — — — — Calculation on the refined model

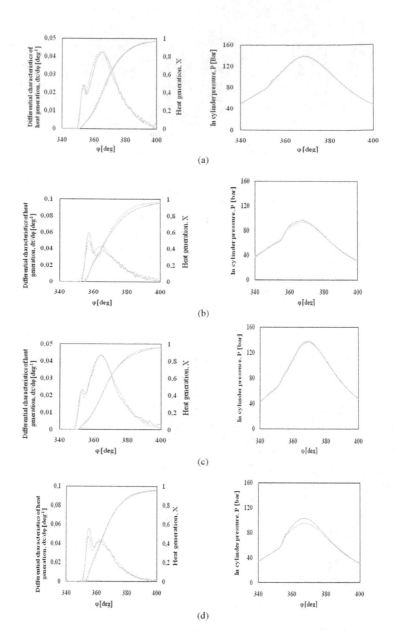

Figure 13. Verification of the model of calculation of heat generation process. EERO: engine speed 2000 rpm, p_e =1.1 MPa (a); engine speed 2000 rpm, p_e =0.56 MPa (b);engine speed 1500 rpm, p_e =1.35 MPa (c);engine speed 1500 rpm, p_e =0.67 MPa (d); ——————— Experiment; — — — — — — Calculation of the refined model

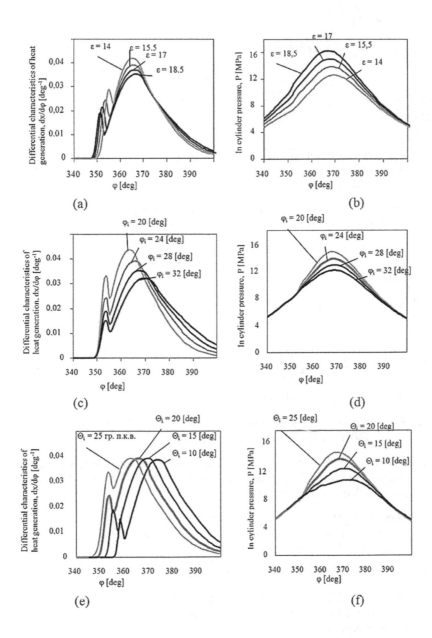

Figure 14. Effect of changing the compression ratio ε (a, b), the injection duration φᵢ (c, d) and injection delay angle Oᵢ (e, f) at the rate of heat generation and pressure in the cylinder of diesel engine

Author details

Andrey Marchenko, Alexandr Osetrov, Oleg Linkov and Dmitry Samoilenko
National Technical University "Kharkiv Polytechnic Institute", Ukraine

References

[1] Priesching, P., Ramusch, G., Ruetz, J. and Tatschl, R., 3D-CFD Modeling of conventional and Alternative Diesel Combustion and Pollutant Formation - A Validation Study. SAE 2007-01-1907 2007.

[2] Tatschl, R., Priesching, P., Ruetz, J. and Kammerdiener, Th., DoE Based CFD Analysis of Diesel Combustion and Pollutant Formation. SAE 2007-24-0048 2007.

[3] Dahlen, L. and Larsson, A., CFD Studies of Combustion and In-Cylinder Soot Trends in a DI Diesel Engine - Comparison to Direct Photography Studies. SAE 2000-01-1889 2000.

[4] Kong, S.-C., Han, Z.Y. and Reitz, RD, The Development and Application of a Diesel Ignition and Combustion Model for Multidimensional Engine Simulations. SAE 9502781995.

[5] Kolade B., Thomas M., Kong SC, "Coupled 1D/3D analysis of fuel injection and diesel engine combustion", SAE International, vol.1, 110 2004.

[6] FIRE v2009 Manual. AVL List GmbH, Graz 2009.

[7] Razleytsev N.F. Modelling and optimization of the combustion process in diesel engines. - Kharkov: Kharkov university press.; 1980.

[8] Kuleshov A.S. The program for calculating and optimizing the internal combustion engine DIESEL-RK. Description of mathematical models, solving optimization problems. M. Bauman. Bauman; 2004.

[9] Hiroyuki Hiroyasu, Toshikazu Kadota and Masataka Arai. Development and Use of a Spray Combustion Modeling to Predict Diesel Engine Efficiency and Pollutant Emissions. JSME1983; 26(214) 576-583.

[10] Wiebe I.I.New about working cicle of engines. Moscow – Sverdlovsk: Mashgiz; 1962.

[11] Watson, N., Pilley, A.D., Marzouk, M. A Combustion Correlation for Diesel Engine Simulation. Diesel Combustion and Emissions . SAE Society of Automotive Engineers 1980; 86 51-63.

[12] Gonchar V.M. Numerical simulation of the working cicle in Diesel. Power machine building; 7 34-35.

[13] Ramos JI. Internal Combustion Engine Modeling. Hemisphere Publishing Corporation; 1989.

[14] Blumberg P., Kummer I.T. Prediction of NO Formation in Spark-Ignited Engines - An Analisys of Methods of Control.Combustion and Flame 1975; 2515-23.

[15] Lino Guzzella and Christopher H. Onder. Introduction to Modeling and Control of IC Engine Systems. Springer; 2007.

[16] ZeldovichY.B., GardenersP.Y., Frank-KamenetskyD.A. Oxidation of nitrogen during combustion. Moscow - Leningrad: AN USSR; 1947.

[17] Semenov V.S. Current problems of the theory of marine diesels. M.: I / O : Mortehinformreklama; 1991.

[18] Varbanets R. The parametric diagnostics of diesel engines SBV6M540 and Pegaso 9156 / R. Varbanets. Aerospace technic and technology 2006; 8 (34) 144-148.

[19] FilipkovskyA.I. Improvement of working process in diesel tipe 32/32 on the basis of physical and mathematical simulation.PhDthesis. HIIT; 1988.

[20] Taldy G.B., Krivobokov A. Numerical Simulation of the combustion process in diesel. Aerospace Engineering and Technology2000; 19158.

[21] LotkoB., LukaninV., Khachiyan, A. Theuseofalternativefuelsininternalcombustionengines. Moscow: MADI (TU); 2000.

[22] Lyshevsky.Theprocessesofatomizationofdieselfuelinjectors. Moskow: Mashgiz; 1963.

[23] Linkov O. Selection and validation of the parameters of mixture formation and combustion of diesel engine operating on alternative fuels. PhDthesis. NTU "KPI"; 2004

[24] MarchenkoA., SemenovV., Sukachev I.I., Linkov O. Simulation of features of the working process in SMD-31 diesel engine running on traditional diesel fuel and methyl esters of rapeseed oil. Aerospace Engineering and Technology2000; 19155-157.

[25] The theory of internal combustion enginesEd. DyachenkoN.H. L.: Mechanical Engineering; 1974. - 552 p.

[26] Zvonov V.A. The toxicity of internal combustion engines. -M.: Mechanical Engineering, 1981.

[27] MarkovV., KislovV., Khvatov V.A. Characteristics of fuel delivery in diesel engines. Moscow: Publishing House of the MSTUBauman; 1997.

[28] ParsadanovI.V. Improving the quality and competitiveness of diesel engines through an integrated fuel efficiency and environmental criteria: the monograph . Kharkov: NTU "KPI"; 2003.

Chapter 8

Physico-Chemical Characteristics of Particulate Emissions from Diesel Engines Fuelled with Waste Cooking Oil Derived Biodiesel and Ultra Low Sulphur Diesel

Raghu Betha, Rajasekhar Balasubramanian and Guenter Engling

Additional information is available at the end of the chapter

1. Introduction

A ground breaking invention in 1893 by Rudolf Diesel (Diesel Engine – named after him) made a mark in the world of internal combustion engines; his engine was the first one to prove that fuel could be ignited without a spark. Since this invention, diesel engines have been widely used in various applications such as automobiles, agriculture, ships, electricity generators, construction equipment etc., all over the world. Diesel engines were proved to be very efficient in terms of delivering the required energy levels for their use at very low operating and maintenance costs when compared to gasoline engines. However, diesel engines now pose a serious threat to human health and adversely impact the urban air quality (Sydbom et al., 2001). Diesel engine exhaust contains a host of harmful substances including airborne particulate matter (PM), carbon soot, toxic metals, polycyclic aromatic hydrocarbons (PAHs), nitrogen oxides which induce ozone formation, carbon monoxide, carbon dioxide, volatile organic compounds and other compounds such as formaldehyde and acrolein (EPA, 2002). Of these pollutants, PM from diesel exhaust is of great concern because of a number of reasons: (1) Diesel engines are known to be the largest source of PM from motor vehicles. Two thirds of PM emitted from mobile sources are from diesel vehicles (EPA, 2002); (2) Human exposure to diesel exhaust particles (DEP) is high as these particles are emitted at ground level unlike that of smoke stacks. United Stated of Environmental Protection Agency (USEPA) reported that 83% of people living in the USA are exposed to concentrated diesel emissions from sources such as highways, heavy industries, construction sites, bus and truck depots etc (EPA, 1999); (3) The freshly emitted DEP includes ultra fine particles (UFPs, aerodynamic diameter (AED) < 100 nm). These particles can bypass the natural defense of respiratory tract and enter deep into the alveolar region of respiratory system from where they could enter into blood stream (Oberdorster, 2001); (4) The particulate-bound soluble organic compounds such as PAHs, nitro-PAHs and transition metals are considered mutagenic or carcinogenic. PM from diesel exhaust is listed as a "likely Carcinogen," citing cancer risk in the range of one in 1000 to one in 100,000 people for each microgram of annual average exposure (EPA, 2002).

Apart from the health impacts, DEP also has potential environmental impacts. Black carbon or soot from diesels affects cloud cover and is a significant contributor to atmospheric warming (Hansen et al., 2000). In view of these concerns and also to meet the current and future regulation standards imposed by local environmental protection agencies, a large number of researchers have conducted research to reduce the diesel particulate levels by various methods such as introducing new engine designs (Guerrassi and Dupraz 1998; Park, et al., 2004), development of particle trap systems and after-treatment devices (Stamatelos 1997), improving the fuel quality (Kaufmann, et al., 1999). Despite these developments over many decades which improved the emission quality of diesels without any doubt, diesel engines still represent a significant source of PM. Holmen and Ayala (2002) reported that the use of particle trap systems reduces total particle number concentration by 10 to 100 times from diesel exhaust, but also found that the use of particle trap-equipped diesel engines may sometimes result in elevated nanoparticle (diameter < 50 nm) emissions. Also, there is an ongoing debate on whether the diesel engine can be modified and improved continuously down to meet the future regulation standards without economic impact. It is believed that developments on engine design needed to meet upcoming regulations would increase the costs and eventually gasoline vehicles could take the place of diesel vehicles (Pischinger, 1996).

As an alternative strategy to improve emission quality of diesels, the option of replacing diesel or petroleum based fuels by renewable bio-fuels is gaining popularity in recent years. Biodiesel (also known as fatty acid alkyl esters), an alternative fuel derived through transesterification process from vegetable oils or animal fats, has received much attention as a result of renewed interest in renewable energy sources for reducing particulate and greenhouse gas (GHG) emissions from diesel engines. In addition, it also helps in alleviating the depletion of fossil fuel reserves (Pahl, 2008; Janauan and Ellis, 2010). Biodiesel is reported to be carbon neutral because of the lower net carbon dioxide production (Ferella et al., 2010; Gunvachi et al., 2007; Carraretto et al., 2004), making it an important fuel source in the era of climate change. Another main driving force for biodiesel is the lower emissions of PM, CO, hydrocarbon, aromatic and polycyclicaromatic compounds (Xue et al., 2011, Atadashi et al., 2010). In view of these advantages, usage of biodiesel is increasing rapidly. This is reflected in government policies such as partial detaxation, investment in production and research of biodiesel by many countries that include USA, Brazil, European Union, South East Asian and other countries all over the world. With these increased awareness and governmental policies of different countries, annual production of biodiesel nearly tripled globally between 2000 and 2005. According to National Biodiesel Board (NBB) in USA alone 460 million gallons of biodiesel were sold in 2007, 700 million gallons in 2008, and 802 million gallons in 2011, showing a tremendous raise from 2 million gallons sold in 2000. This move from fossil fuels to bio-fuels as a power source is also caused by the economic consequences due to stringent regulations ((EN-590, 2004) in Europe and (ASTM-D-975, 2006) in USA) imposed on the fuels used in transportation. The notable restriction includes reducing sulfur content in the fuel which increased the investment cost of oil companies and the final fuel price which drive the nations to develop their own reserves indigenously and decrease their dependence on Middle East countries for fuel.

Biodiesel is compatible with conventional petroleum based-diesel and can be completely blended with diesel in any proportion. The chemical composition and several properties of biodiesel make it an attractive option over traditional diesel. Biodiesel has higher cetane number, lubricity, combustion efficiency, biodegradability and lower sulfur and aromatic content (Fazal et al., 2011; Demirbas, 2008). In contrast, there are also unfavorable properties in biodiesel such as being more prone to oxidation, lower heating value, and higher cloud and pour points (Szulczyk and McCarl, 2010). Majority of biodiesel is being produced from soybean, rapeseed, and palm oils. Even though most of the biodiesel feedstock is renewable, competition with food supply has become a serious concern recently because certain feedstocks appear to be edible oils (Januan and Ellis, 2010; Mercer-Blackman et al., 2007). Therefore, alternative feedstocks such as non-edible oils, algae oils, and waste oils have arisen to prominence in recent years. Biodiesel produced from transesterification of waste cooking oil (WCO) is one of the most attractive automotive fuels to be used in place of petroleum diesel because of the added advantages over other types of biodiesel. WCO reuse eliminates the need for disposal, thus alleviating the environment and human health issues associated with waste oil disposal (Giracol et al., 2011). The lower cost of WCO feedstock can also help to make biodiesel competitive in price with conventional diesel (Meng et al., 2008). Many studies have been initiated to investigate the impacts of biodiesel made from several feedstocks including WCO on particulate emissions as compared to diesel fuel (Chung et al., 2008; Lin et al., 2011a; Lapuerta et al., 2008, Turrio-Baldassari et al., 2004; Durbin et al., 2007). An apparent decrease in PM emissions with the biodiesel content can be considered as an almost unanimous trend (Lapuerta et al.,2008). Most of the research on emissions from biodiesel is targeted towards physical properties of PM such as particulate mass, number concentrations and their size distributions. Apparently very little information is available pertaining to the health and environmental impacts of particulate emissions of biodiesel due to paucity of data on their chemical composition.

2. Phyiscal characterization of PM emitted from WCOB and ULSD

Conventional regulatory procedure involving dilution tunnel sampling and filter collection proved to be satisfactory for collection of PM from diesel engines a decade ago. However, in the current scenario, the low emission rate (~1mg/km) of PM from modern-era diesel engines places difficulties in sampling through traditional procedures because of their high detection limits. Nor is this traditional method agreeable to recent regulations aimed at in-use emissions monitoring or to the vast variety of off road applications. A workshop was organized by Coordinating Research Council (CRC) in 2002 to discuss possible changes to measurement of DPM, and it is proposed that particle number-based methods potentially allow detection at very low levels with consistency provided formation of nucleation mode particles can be avoided (CRC, 2002). However, it is a difficult task to establish particle number-based standards and a standard methodology for measurement because of the sensitivity in detection, and great variability of nano-sized particles in engine exhaust. The European PMP (Particle measurement Program) and many other organizations including USEPA are working towards improving the methodology for measurement of solid particles to supplement the traditional mass method. A well-designed dilution tunnel satisfying the above requirements and reducing losses is the first step for this purpose.

Typically, DEP are agglomerates of many primary spherical particles of about 15-40 nm diameter. Airborne particles differ in size, composition, solubility and therefore also in their toxicological properties. It is a well-established fact that the current standards on diesel engine emissions not only improved the engine technology but also the fuel quality. These modern day engines emit particles of very low diameter (Su et al., 2004). Most of the particle mass exists in the accumulation mode in the diameter range of 0.1-0.3 μm (Kittelson, 1998). A large part of UFPs go unnoticed when only mass concentration is used as a metric. UFPs contribute a small fraction to the mass concentration of ambient aerosol particles, but may contribute disproportionately to their toxicity because of their high number concentration and surface area, high deposition efficiency in the pulmonary region, and high propensity to penetrate the epithelium (Donaldson et al., 2000).

2.1. Design of dilution tunnel

A dilution tunnel is a closed and controlled chamber where hot exhaust from engines, industrial stacks etc is mixed with dilution gas (usually ambient air) prior to sampling. Dilution sampling was originally used to characterize fine particle emissions from combustion sources because it simulates the rapid cooling and dilution that occurs as exhaust mixes with the atmosphere. Several researchers have developed dilution tunnels over decades for this purpose; one of the popular dilution sampling systems to simulate atmospheric conditions is CALTECH design (Hildemann et al., 1989). However dilution tunnels can also be used to freeze the size distribution by proper design criteria to avoid unwanted nucleation, condensation and coagulation. These dilution tunnels are developed for consistent measurement of particle number concentrations (PNC) from diesel engines.

The design of the sampling and dilution system determines largely what is measured later. Burtscher (2005) suggested in his review paper that the solid nanoparticles from the exhaust are closely related to health impacts and thus solid fractions of the exhaust should be separated from the volatile fraction and be studied for better understanding of health effects. Also, Kittelson (1998) identified that nucleation and coagulation of volatile fraction changes dramatically during dilution and sampling making it difficult to design a standard. Thus, one of the major issues in dilution sampling of engine exhaust is to decrease or eliminate nucleation processes. Typical dilution tunnels for particualte sampling from engines are de-signed to meet the following requirements.

a. To reduce the particle concentration in raw exhaust to a concentration that can be han-dled by the measurement system;

b. To reduce the temperatures to an adequate value usually close to ambient temperature;

c. To control the nucleation/condensation processes;

d. To reduce the losses of particulate matter during dilution and sampling.

a. Reducing the PM concentration

The PM concentration in raw exhaust varies with engine model, design, applied load and other parameters. A rough estimate of the total number concentrations of diesel engines is in the range of 10^8~10^9 cm^{-3}. The present day sampling equipment and monitoring devices are designed to capture/monitor the particles in nanoscale that are abundantly present in to-day's diesel exhaust. These sampling instruments are highly sensitive, fragile and are relia-ble. Heavy loading of PM can easily disrupt the configuration, damage crucial systems leading to either inaccurate data, or render the equipment to be useless. So, it is required that the concentrations be brought down well below the instrument's maximum capacity.

b. Reduction in exhaust temperature

A typical temperature of raw exhaust is in the range of 200~400°C, and such high tempera-tures can damage the charger columns and sensitive electrode plates used for detecting nanoparticles. Also, collection of PM on filter media from hot raw exhaust can alter the chemical composition of PM by inducing chemical reaction between the collected particles and filter media. Current regulation, for example the one used by the USEPA, requires that, PM be collected on filter media after the exhaust has been diluted and cooled to a tempera-ture below 52°C (CFR, 2001).

c. Reduction in nucleation /condensation and coagulation

Engine exhaust contains both gaseous phase and particle phase pollutants. Because of lower volatility and saturation coefficient some organic compounds and other precursor gases such as sulfur dioxide either condense onto the pre-existing particle surface altering their size (diameters), or nucleate to form new particles affecting number concentration. Also, particles can coagulate during dilution changing both the diameter and number of particles. Impact of these processes is very uncertain as they change dramatically during dilution and sampling (Kittelson, 1998) and thus makes comparisons from different sources difficult. It is therefore recommended that the dilution tunnels be designed to minimize or eliminate nu-cleation, condensation and coagulation (Burtscher, 2005).

d. Reducing losses during sampling

Particles are lost during transfer of exhaust from a tail pipe to sampling instruments and during dilution due to particle – wall interactions. These losses include mechanical losses such as inertial impaction, gravitational settling, electrostatic deposition, and due to diffusion (Kittelson and Johnson 1991). Apart from mechanical losses particles are also lost due to thermophoretic deposition. These losses can impact the particle number concentration and size distribution.

The primary design objective of the dilution tunnel is to make sure that what is released at tail pipe is measured at sampling instruments. In other words, it is designed to minimize/ eliminate dilution artifacts, reduce losses, avoid nucleation to preserve the number and size distribution of particles as it is emitted from tail pipe. Before describing the actual design, it is useful to understand the theoretical basis underlying the design. The following sections 2.1.1 and 2.1.2 describe the important mechanisms that play a key role in altering the physical and chemical properties of particles during dilution and the measures taken to prevent such changes. Section 2.1.3 presents actual design of the dilution tunnel.

2.1.1. Nucleation, condensation an coagulation

Concentration of an inert species (C_i) when diluted is given by (Kerminen and Wexler 1995).

$$C_i - C_{i,A} = f(C_{i,E} - C_{i,A})$$

(1)

where, $C_{i,E}$ and $C_{i,A}$ are concentrations in the exhaust and ambient air respectively, and f is dilution factor. Dilution factor simply means when the exhaust dilutes, a small parcel of air contains a certain fraction of the original exhaust and the remaining fraction is ambient air. That fraction of original exhaust in the air parcel is the dilution factor (f).

Temperature of an air parcel also changes in similar manner.

$$T - T_A = f(T_E - T_A)$$

(2)

where, T_E and T_A are the exhaust and ambient temperature, respectively.

a.   Nucleation

Nucleation and condensation go hand in hand. Nucleation of nanoparticles from the diesel exhaust takes place as the exhaust cools during the dilution process (Abdul-Khalek et al., 1999). When partial pressure is much higher than vapor pressure of nucleating species, they undergo phase transformation. At the same time, due to their low volatility and the existence of large surface area of particles, sulfuric acid and many organic compounds can also condense quickly on the particles. However, Zhang and Wexler (2002) reported that nucleation favors compounds with both low volatility and low-molar volume in the condensed phase. High carbon number organics which usually possess very low vapor pressure, but significantly large condensed-phase molar volume are less likely to be the nucleating species than sulfuric acid (Zhang and Wexler 2004). Although it is still not clear about the nucleation mechanism from SO_2, it is widely believed that nucleation occurs through binary nucleation of water-sulfuric acid system (Kulmala et al., 1990) or ternary nucleation of Water-

H_2SO_4-NH_3 system (Korhonen et al., 1999). Precursor gas, sulfur dioxide, is oxidized to form trioxide which gets converted to sulfate with water vapor.

$$SO_2(g) + \tfrac{1}{2}O_2(g) \rightarrow SO_3(g) \tag{3}$$

$$SO_3(g) + H_2O(g) \rightarrow H_2SO_4(g) \tag{4}$$

The critical value (C_{crit}) for the gas phase concentration of sulfuric acid required for binary nucleation to take place is given by the following formula (Jackervoirol and Mirabel, 1989).

$$C_{crit} = 0.16\exp(0.1T - 3.5rh - 27.7) \tag{5}$$

where C_{crit} is in $\mu g/m^3$, T is Temperature in Kelvin, rh is scaled between 0-1. When the critical ratio $H_2SO_4(g)/C_{crit}$ becomes greater than 1, nucleation (gas-particle conversion) occurs instantaneously, giving birth to fresh nuclei in another log-normal distribution; these are called dilution-induced nuclei mode particles. From eq (5), it is clear that critical concentration is a function of temperature. When the temperature of diluted exhaust is high, the critical concentration required for nucleation tends to increase exponentially. Thus, it is recommended to dilute the hot exhaust with heated air rather than the conventional way of diluting the exhaust with air at ambient temperature. Also, Zhang and Wexler (2004) studied the viability of sulfuric acid induced nucleation and the coupling effect of condensation and nucleation by simulating various cases and reported two important findings: (1) As surface area of combustion induced particles increases, the critical ratio drops and in extreme cases nucleation is totally quenched. This is not to say that supersaturated mixture is not formed, but the sulfuric acid is condensed onto the surface area of particles; (2) Extremely rapid dilution leads to super-saturation of nucleating species and thus nucleation. The possibility of nucleation becomes lower if the dilution is smooth.

b.   Condensation

The vapor pressure for volatile compounds is proportional to its temperature. Vapor pressure is related to dilution factor by (Zhang and Wexler 2004)

$$p^0 \, \alpha \exp\left(-\frac{\Delta H}{Rf T_E}\right) \tag{6}$$

Close to tail pipe, the temperature of exhaust far exceeds the ambient temperature, and therefore equation (2) reduces to (T α f). As a result, a decrease in temperature leads to an exponential decrease in vapor pressures, and the highly super-saturated vapors could make the time scale of condensation very short as 0.1 sec. Another factor important for condensation is particle surface area; by reducing the available particle surface area, condensation can be quenched.

c. Coagulation

Coagulation of particles during dilution occurs either due to turbulent shear, or due to Brownian coagulation. Zhang and Wexler (2004) evaluated coagulation through turbulent shear which has a time scale of $\tau_{ts} = \rho_p \big/ \big(48\, \hat{m}_p \sqrt{\varepsilon_k / 120 v}$ and found that τ_{ts} for turbulent shear is approximately 10^{18} sec, because these tiny particles have such small cross sectional area that shear is insufficient to bring them together. The only mechanism is through Brownian coagulation and it is very insignificant when compared to other mechanisms.

2.1.2. Particle-wall interaction

When the engine exhaust is sampled and passed through dilution tunnel and sampling tubes, particles are lost due to deposition on sampling surfaces. Particles are deposited through several ways which include mechanical (inertial, gravitational, electrostatic, diffusion) and thermal (thermophoretic losses). In their review of variability in particle emission measurements during heavy duty transient tests, Kittelson and Johnson (1991) discussed the impact of exhaust system temperatures on particle measurement and provided recommendations to minimize the effects on aerosol sampling. They calculated losses during the heavy-duty transient test for a typical test facility and found that the majority of particle loss (5%) is due to thermophoresis, whereas inertial, gravitational, electrostatic, and diffusion depositions put together resulted in a loss of 0.2%. Thermophoresis is a physical phenomenon in which particles, subjected to a temperature gradient, move from high- to low temperature zones. A temperature gradient is established during sampling and the dilution between exhaust and the sampling surfaces due to difference in their temperatures. This gradient results in thermophoretic deposition of particles on sampling and dilution system surfaces. Eventually, these deposits are reentrained in the exhaust stream, or cause fouling of sampling surfaces. Reentrainment is unpredictable, and increases variability in mass measurements because of the increase in the number of coarse particles. These particles are not necessarily representative of diesel aerosol and make aerosol size distribution measurements more difficult. To avoid large differences in temperature between lines and exhaust gas, sampling lines should be fully insulated and kept to optimum size to reduce the residence time of exhaust in the sampling lines. Short sampling lines also reduce gravitational and diffusional losses.

2.1.3. Dilution tunnel

Figure 1 shows the schematic of a typical dilution tunnel used for particulate sampling from diesel engines. The engine exhaust was sampled through a sampling probe inserted into the main exhaust stream. The exhaust was then directed into a dilution tunnel, where the hot exhaust was mixed with a stream of pressurized, particle free, dry air in two stages using. In the first stage, the exhaust was transferred into the primary dilution tunnel, where the hot exhaust was mixed with a stream of pressurized particle free dry air preheated to a temperature of close to the temperature of the exhaust, to avoid particle nucleation, condensation, etc. The primary dilution tunnel was also heated to a temperature close to engine exhaust temperature to avoid thermophoretic deposition of particles onto the walls of the dilution tunnel. In stage 2, some portion of the diluted exhaust from the primary dilution tunnel was transferred to the secondary dilution tunnel. During the primary dilution, vapor pressures of volatile compounds decreased, allowing the secondary dilution with cold dilution air without condensing the volatile components. At this stage, the exhaust was mixed with dilution air at ambient temperature to bring down the temperature of the hot exhaust to ambient temperature for sampling. The completely diluted exhaust was then directed to particle measuring instruments. The dilution ratio was calculated by measuring the CO_2 concentrations in the raw exhaust, dilution air and diluted exhaust.

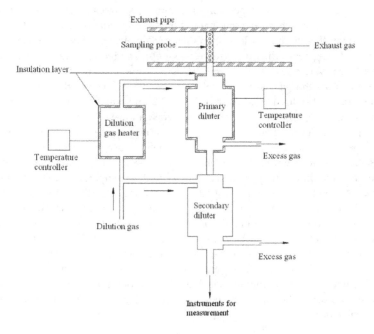

Figure 1. Schematic of a typical dilution tunnel.

A two stage dilution is adopted, firstly to prevent rapid dilution, a dilution is not achieved instantaneously but gradually and secondly, to dilute the exhaust in two different environments to reduce nucleation/condensation. In the primary dilution tunnel, since the exhaust is mixed with pre-heated air, time required for condensation is high comparable to the residence time of the exhaust in dilution tunnel (1~2 sec). In the secondary dilution tunnel, since a small sample from the already diluted exhaust is drawn, the available surface area for condensation is reduced by many times thus condensation is avoided in both stages of dilution.

2.2. Particulate mass concentration

Biodiesel (BD) and blends of ultra low sulfur diesel (ULSD) – BD had a significant effect on particulate emissions. Studies were conducted extensively on PM mass emissions from engine fueled with BD. Although some authors have reported an increase in PM emissions relative to diesel (Durbin et al., 2000; Munack et al., 2001; Alfuso et al., 1993), a large number of studies have confirmed a noticeable decrease in particulate mass while using BD (Lapeurta et al., 2008). However, the reported reductions varied very much depending upon engine conditions, experimental set up, fuel used, engine fuel system and other factors. Several studies reported reductions in the range of around 40 to 50% (Krahl et al., 1996, Lapeurta et al., 2002; Bagley et al., 1998). However, reductions as high as 70 to 90% were also reported by few studies (Canakci and Van Gerpen, 2001; Camden Australia 2005; Kado, 2003; Kalligeros et al., 2003). $PM_{2.5}$ (AED ≤ 2.5 µm) concentrations for ULSD and WCOB blends were reported by Betha and Balasubramanian, 2011a. They observed that with an increase in percentage of biodiesel in the fuel mixture, particle mass was reduced for B50, and for WCOB (B100) for all loading conditions of engine. For a particular fuel blend, the PM mass increased with load. Lapuerta et al. (2008) reported that many previous studies observed PM reductions in the range of 40-70% when biodiesel was used. The percentage reduction of $PM_{2.5}$ reported by Betha and Balasubramanian, 2011a (~35% at full load) was slightly lower than the range reported by Lapuerta et al. (2008). This is mainly because most of the studies, as reported by Lapuerta et al. (2008) in their review paper, compared biodiesel to conventional low sulfur diesel (sulfur content < 500 ppm) whereas in the study conducted by Betha an Balasubramanian 2011a the comparison of $PM_{2.5}$ emissions was made with ULSD (sulfur content < 15 ppm). The reduction in PM emissions when using BD can be attributed to the following reasons: (1) the absence of aromatics (Knothe et al., 2006), which are considered as soot precursors, in biodiesel reduces the amount of PM formed during combustion; (2) the higher oxygen content in BD tends to enhance the combustion process resulting in lower particulate emissions; and (3) Finally, the presence of unsaturated fatty acids in BD leads to more complete combustion processes. Unsaturated fatty acids have lower boiling points than diesel, and they can evaporate faster in the combustion chamber than diesel (Song and Zhang, 2008). In addition, the higher viscosity and density of BD compared to ULSD can lead to an increase in the injection pressure. Likewise, higher bulk modulus of compressibility of vegetable oils and their methyl esters can lead to advanced injection timing (Boehman et al., 2004) while using BD. As a result, the BD fuel enters the combustion chamber relatively quicker compared to ULSD (Lapuerta et al., (2008). This advancement in combustion process while using BD increases the residence time of soot particles in the combustion chamber, and thus they undergo further oxidation (Cardone et al., 2002) leading to reduction in PM emissions.

2.3. Particle number concentrations and size distributions

New engine designs and emission control devices reduced particulate mass emission drastically allowing the engines to operate below the emission level standards. However, the concerns about UFPs and nanoparticles (AED < 50 nm) which can contribute to human health effects (Oberdiester et al., 2001, Nel, 2005) significantly raised a serious concern to develop new ambient standards in terms of particle number rather than mass (Burtscher, 2004). Therefore, particulate number concentrations and size distributions are increasingly studied in comparative studies of particulate emissions from diesel and BD (Kittelson, 1998; Lapeurta et al., 2008; Zhu et al., 2010, Di et al., 2009a and 2009b; Di et al., 2008; Burtscher, 2004). In the literature, both an increase and decrease in the total particle number concentrations were reported when using BD. Di et al. (2009b) observed that the total particle number increased 1.5 – 2.5 times when using WCOB compared to diesel depending on the engine load. Similar increments (1.35 – 2.4 times) were observed by them in another study using a direct injection diesel engine. On the other hand, Lapuerta et al. (2007) tested two differently stressed WCOB and observed a reduction (~3 times) in total particle number concentrations (PNC) compared to diesel. Although both of them used WCO in their study, contradictory trends were reported. Studies on other types of BD have also reported contradictory results. A summary of PNC of ULSD and WCOB emissions during different engine operating conditions is provided in Table 1 (Betha et al., 2011a) and their size distributions are shown in Figure 2 (Betha et al., 2011a). In contrast to PM mass emissions, PNC decreased with an increase in load for all the fuels (shown in Table 1). At higher WCOB-ULSD blend ratios, the percentage decrease in PNC with increasing load was relatively small compared to ULSD. For ULSD, total PNC decreased by 26% at full load when compared to idle, or no load conditions whereas, for biodiesel, it decreased by only around 9%.

Engine Load (%)	Total particle number concentration(# cm^{-3})			
	ULSD	B20	B50	B100
0	1.14×10^7	1.05×10^7	9.57×10^6	8.98×10^6
30%	9.82×10^6	8.99×10^6	9.19×10^6	8.58×10^6
70%	9.00×10^6	8.91×10^6	8.87×10^6	8.40×10^6
100%	8.46×10^6	8.34×10^6	8.31×10^6	8.15×10^6

Table 1. Total PNC for ULSD and WCOB (B100) blends at various loads (Betha et al., 2011a)

It is expected that in diesel engines, particle counts would increase with an increase in load. However, a decrease in PNC with increasing load was observed in this study. Chung et al. (2008) also observed a reduction in PNC at higher loads in their study using a Yanmar backup generator similar to the one used in this study. This observed decrease in PNC was probably due to the transfer of particles from nucleation to accumulation or coarser mode at higher loads. It was observed that at higher loads, nucleation mode particles which contribute to a major fraction of total number decreased and accumulation mode particles increased. As a result, there was an overall decrease in total PNC. This shift in particles from the nucleation mode to the accumulation mode was evident from the PSD shown in Figure 2 (Betha et al., 2011a).

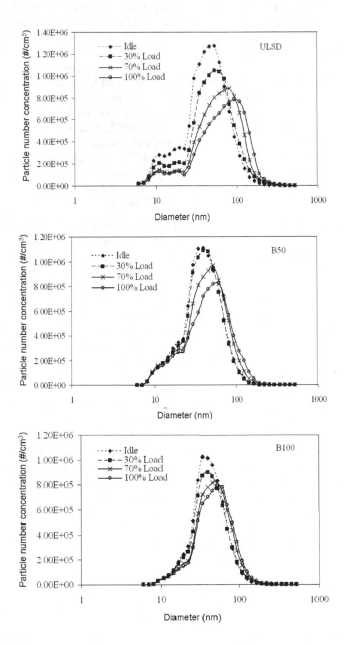

Figure 2. Particle size distributions of emissions at different loads for ULSD, B50, WCOB (B100) (Betha et al., 2011a)

From the PSD shown in Figure 2, it can be observed that with the increase in load, particle peak diameters increased for all fuels. The magnitude of the increase in particle peak diameters at higher loads was greater for the ULSD than the biodiesel. For ULSD, the peak diameter increased from 52.3 nm at idle mode to 93 nm at full load and for biodiesel (B100) peak diameter increased from 34 nm at idle to 52.3 nm at full load. However, in their study on particle emissions from stationary diesel engines, Di et al.(2010) and Zhu et al. (2010) reported an increase in total PNC when engine load was increased. Although there is no definitive explanation for the difference in particle emission trends observed by the above-mentioned studies and trends shown in Figure 2 as well as the study by Chung et al. (2008), a possible reason can be the differences in engine capacities and operating conditions. Both Di et al. (2010) and Zhu et al. (2010) conducted their studies on diesel engines which had much larger capacity and power (~4000 cc and 88 kW). However, the stationary diesel generator used for the the shown results (Table 2 and Figure 2) (296 cc and 4.5 kW) and the one used by Chung et al. (2008) (4.8 kW) have lower capacity and power.

The relative number percentage of nucleation (diameter < 50 nm), accumulation (50-100 nm) and fine particles (diameter > 100 nm) emitted from diesel and biodiesel (B100) fuels at various loads is shown in Figure 3. Biodiesel had relatively a higher fraction of nucleation mode particles ranging from 55% to 70 % at different loads when compared to diesel (35% – 60%). A decrease in the nuclei mode particles (diameter < 50 nm) and an increase in accumulation and fine particles (diameter > 100 nm) were observed with an increase in load for both the fuels. For the stationary engine running with ULSD at full load, nucleation (32%), accumulation (40%) and fine particles (28%) shared almost a similar fraction of particles to the total PNC. However, for BD, nucleation (51%) and accumulation mode particles (43%) were major contributors to the total number concentrations. The fraction of accumulation mode particles increased from 30% during idle mode to 40% at full load in the case of ULSD, and a similar increase was observed for biodiesel as well (from 30% at idle mode to 43% at full load). This observation implies that diesel engines emit more accumulation particles at higher loads. At higher loads, more fuel is injected into the combustion chamber to generate additional torque needed and also the residence time for the particles in the combustion chamber decreases relatively. Therefore, the oxidation of particulate soot tends to be reduced, leading to the release of a large fraction of accumulation and fine particles. In the case of biodiesel, the inherent oxygen in the fuel improves the oxidation of soot. Therefore, the percentage increase of fine particles is relatively less for BD.

3. Chemical properties of particulate emissions

Since diesel engines are one of the most significant air pollutant sources in urban areas (Cass, 1998), chemical composition of diesel exhaust has been widely investigated. The chemical profile of PM plays a crucial role in health and environmental impacts. Variations in the chemical composition of aerosols alter their hygroscopicity and can lead to changes in the cloud-active fraction of the aerosols, or cloud condensation nuclei (CCN) number concentration (Ward et al., 2010). Some carcinogenic and toxic chemical compounds present in DEP when biologically available can affect human health. Diesel engine emissions consist of a wide range of organic and inorganic compounds in gaseous as well as particulate phases (Bünger et al., 2006). Concentrations of most particle-bound chemical constituents depend on the type of engine, engine load, fuel and lubrication oil properties (Dwivedi et al., 2006). Large surface area of DEP enables adsorption of organics and inorganic compounds from the combustion process and/or the adsorption of additional compounds during transport in the ambient air. DEP consists mainly of elemental carbon (EC) (75%), organic carbon (OC) (19%) and small amounts of sulfates, nitrates (1%) and metals & Elements (4%) (Figure 4) (EPA, 2002a).

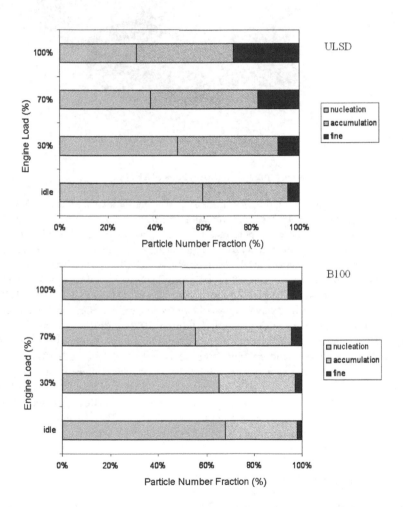

Figure 3. Fractionation of particle emitted from ULSD WCOB (B100) for various loads (Betha et al., 2011a).

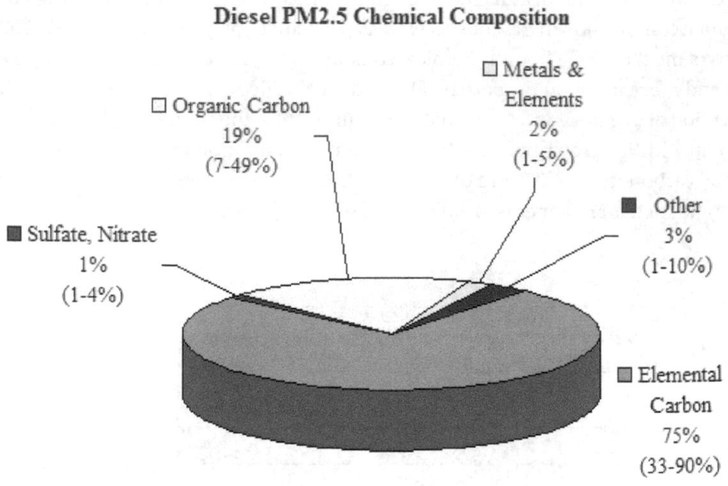

Diesel PM2.5 Chemical Composition

Figure 4. Typical chemical compositions for diesel particulate matter (PM$_{2.5}$) (EPA, 2002a)

3.1. Particle-bound polycyclic aromatic hydrocarbons (PAHs)

Particulate-bound organic compounds, especially PAHs, are highly carcinogenic. PAHs and their derivatives (nitro-PAHs) together comprise less than 1% of the mass of DEP (EPA, 2002a). The emissions of these compounds are comprehensively studied for diesel engines fuelled with diesel and BD (Jung et al., 2006; Bagley et al., 1998; Correa et al., 2006; Karavalakis et al., 2009; Turro-Baldassari et al., 2004; Zou and Atkinson, 2008; Karavalakis et al., 2010). A majority of studies have found a significant decrease in PAHs emissions with BD compared to that with diesel (Bagley et al., 1998; Correa et al., 2006; Karavalakis et al., 2009). However, a couple of studies have indicated only statistically insignificant reduction in PAHs (Turro-Baldassari et al., 2004; Zou and Atkinson, 2008) when BD is used. The reduction in PAH emission may be attributed to the presence of excess oxygen in BD and the absence of aromatic and polycyclic aromatic compounds in the fuel. One study (Karavalakis et al., 2010) was found in literature reporting higher PAHs emissions when using BD. Karavalakis et al. (2010) tested BD made from soybean oil and used frying oil. They found lower PAH emissions with BD made from soybean oil. However, BD made from used frying oil emitted more PAH compounds compared to those from diesel. They attributed the increase in PAHs to dimers, trimers, polymerization products, and cyclic acids present in the biodiesel made from used frying oil.

3.2. Particulate-bound elements and trace metals

Studies on particulate-bound metals emitted from BD combustion are not as comprehensive as PAHs, despite a strong correlation between human health risk and particulate-bound metals (Hu et al., 2008; Verma et al., 2010). Very few studies (Dwivedi et al., 2006; Cheung et al., 2011) were found in literature investigating particulate-bound metals in BD. Dwivedi et al. (2006) conducted a comparative assessment and characterization of particulate-bound trace elemental emissions from diesel and rice bran oil derived BD. Elements such as Cr, Ni, Pb, Cd, Na, Al, Mg, and Fe were investigated. The authors observed that concentration of metals such as Cr, Fe, Al, Zn, Mg increased while others (Pb, Cd, Na, Ni) were reduced with the usage of B20 (20% BD). Cheung et al. (2011) also observed higher concentrations of Fe, Zn, Mg when using BD. However, Cr, Al, Pb, Cd, Na, Ni were lower in BD emission compared to diesel. In both the studies, the particulate-bound elemental concentrations were mainly attributed to the fuel and lubricating oil composition apart from engine wear. Metals and elements that were found higher in fuel were also found higher in the emissions. Since the particulate-bound elements largely depend on fuel quality and composition apart from engine wear, their concentration in the exhaust is expected to vary with feedstock of BD. In another study Betha et al. (2011b) investigated the particulate bound elements from WCOB (B100), ULSD and their blend (B50). They observed that particulate emissions were reduced with the usage of WCOB. However, most of the elements which are known to be toxic such as Zn, Cr, Ni were very high in the WCOB exhaust compared to ULSD. Elements such as As, Co, Al, Mn were found to be in higher levels in ULSD. Elements such as Cr, Cu, Fe, Ba, Zn, Mg, Ni, and K were found in higher concentrations in B100 compared to ULSD. Similar findings were reported by Dwivedi et al. (2006). They found that concentrations of Zn, Fe, and Cr were higher in biodiesel compared to those in diesel. The higher concentrations of elements, especially Cu, Fe, Zn, in B100 used in this study can be attributed to the raw material from which biodiesel was prepared. BD used in this study was derived from WCO generated in restaurants and food courts. The oil has been used for cooking and frying of various food products. Elements such as Cu, Fe, Zn and Mn were found in vegetables, (Kawashima and Soares, 2003) Cu, Fe, and Zn in meat (Lombardi-Boccia et al., 2005) and Cu, Zn and Cd in fish (Atta et al., 1997). These elements might have been released into the oil during cooking and therefore, the concentrations of Fe, Zn, Cu, Mn in BD were found to be significantly high. In addition, these elements can also be leached out from the cooking utensils due to heating (Kuligowski and Halperin, 1992).

4. Estimation of health risk due to particulate emissions

Human health risk assessment was conducted based on the mean concentrations of particulate-bound elements determined through the experimental study. Health risk assessment is especially useful in understanding the health hazard associated with inhalation exposure to PM emitted from B100 compared to that of ULSD. The details and steps involved in health risk assessment are described in detail elsewhere (See and Balasubramanian, 2006). Briefly, it involves four important steps (NRC, 1983) as described below.

Hazard identification – elements which have known toxicity values are considered. Al, Cr, and Mn induce non-carcinogenic effects while As, Cd, Cr, Ni and Co induce carcinogenic health effects.

Exposure assessment –This involves estimation of chronic daily intake (CDI) of these elements calculated from the following equations.

$$CDI(mg\ kg^{-1}day^{-1}) = \frac{Total\ dose\ (TD,mg\ m^{-3}) \times inhalation\ rate\ (IR,\ m^3\ day^{-1})}{Body\ weight\ (BW,kg)} \qquad (7)$$

$$TD = C \times E \qquad (8)$$

where C is concentration of pollutant and E is deposition fraction of particles by size given by (Volckens and Leith, 2003),

$$E = -0.081 + 0.23\ln (D_p)^2 + 0.23\ sqrt\ (D_p) \qquad (9)$$

where Dp is the diameter of particles. In this study, $PM_{2.5}$ (Aerodynamic diameter $\leq 2.5\ \mu m$) was used i.e. Dp is 2.5 µm. IR is typically assumed to be 20 m^3 day^{-1} and BW to be 70 kg for adults. As for children, the IR and BW are assumed to be 10 m^3 day^{-1} and 15 kg, respectively.

(3) Dose-response assessment- It is the probability of health effects according to the dose of pollutant of concern. Assuming only inhalation as the major exposure route, the reference dose (RfD, mg kg^{-1}, day^{-1}) for toxic elements that are non-carcinogenic was calculated from reference concentrations (RfC, mg/m^3) provided by USEPA. Likewise, for carcinogenic elements the inhalation slope factor (SF, mg^{-1} kg day) was calculated from inhalation unit risk values (IUR, mg^{-1} m^3) provided by USEPA.

(4) Risk characterization or estimation of health risk - was calculated based on the exposure and dose–response assessments. For non-carcinogenic metals, it is indicated by (United States Department of Energy, 1999):

$$Hazard\ Quotient\ (HQ) = CDI/RfD \qquad (10)$$

For carcinogenic metals, total carcinogenic risk is estimated in terms of excess life time cancer risk (ELCR) given by: (United States Department of Energy, 1999).

$$ELCR = CDI \times SF \qquad (11)$$

The human health risk assessment was carried out to quantify the risk associated with the particulate-bound metals emitted from ULSD and WCOB at full load from a stationary engine for illustration.

The pertinent information of the TD and RfD, HQ, inhalation SF and ELCR for adults and children are shown in Tables 2 and 3. The concentrations of metals used for this illustration are adopted from Betha et al. (2011b) (shown in Table 4) and CDI is calculated using Eqns (7) – (9). The concentrations reported in Betha et al., 2011b represent those emitted from the raw engine exhaust. However, in reality the engine exhaust is diluted by ambient air once it is released to the atmosphere. The dilution factor is typically 1000 times when exhaust is released to the ambient atmosphere on road conditions (Zielinska, 2005). The mean concentrations of elements were divided by 1000 to be used in health risk calculations. The deposition efficiency E for particle with 2.5 μm is nearly 0.70. The CDI is calculated by first estimating the TD of the each element (Tables 2 and 3).

ULSD

Metals	CDI (mg kg^{-1} day^{-1})	RfD (mg kg^{-1} day^{-1})	HQ	SF (mg kg^{-1} day)	ELCR
Non carcinogenic metals					
Al	9.84 x 10^{-6}	1.43 x 10^{-3}	6.88 x 10^{-3}		
Cr	1.22 x 10^{-6}	2.86 x 10^{-5}	4.26 x 10^{-2}		
Mn	9.18 x 10^{-8}	1.43 x 10^{-5}	6.42 x 10^{-3}		
Carcinogenic metals					
As	2.51 x 10^{-8}			15.1	3.79 x 10^{-7}
Cd	3.79 x 10^{-9}			6.3	2.39 x 10^{-8}
Cr	1.22 x 10^{-6}			4.2	5.12 x 10^{-6}
Ni	8.02 x 10^{-8}			84	6.74 x 10^{-6}
			$\sum$=5.59 x10^{-2}		$\sum$ =12.3 x10^{-6}

WCOB

Metals	CDI (mg kg^{-1} day^{-1})	RfD (mg kg^{-1} day^{-1})	HQ	SF (mg kg^{-1} day)	ELCR
Non carcinogenic metals					
Al	6.42 x 10^{-6}	1.43 x 10^{-3}	4.49 x 10^{-3}		
Cr	2.18 x 10^{-6}	2.86 x 10^{-5}	7.63 x 10^{-2}		
Mn	3.85 x 10^{-8}	1.43 x 10^{-5}	2.69 x 10^{-3}		
Carcinogenic metals					
As	9.57 x 10^{-9}			15.1	1.45 x 10^{-7}
Cd	2.31 x 10^{-9}			6.3	1.45 x 10^{-8}
Cr	2.18 x 10^{-6}			4.2	9.16 x 10^{-6}
Ni	8.72 x 10^{-8}			84	7.33 x 10^{-6}
			$\sum$=8.34x 10^{-2}		$\sum$=16.6 x 10^{-6}

Table 2. Estimation of human health risk in adults due to particulate bound elements from PM$_{2.5}$ emitted from WCOB and ULSD (Betha et al., 2011b)

ULSD

Metals	CDI (mg kg^{-1}day^{-1})	RfD (mg kg^{-1}day^{-1})	HQ	SF (mg kg^{-1}day)	ELCR
Non carcinogenic metals					
Al	2.30×10^{-5}	1.43×10^{-3}	1.61×10^{-2}		
Cr	2.85×10^{-6}	2.86×10^{-5}	9.95×10^{-2}		
Mn	2.14×10^{-7}	1.43×10^{-5}	1.50×10^{-2}		
Carcinogenic metals					
As	5.85×10^{-8}			15.1	8.84×10^{-7}
Cd	8.85×10^{-9}			6.3	5.58×10^{-8}
Cr	2.85×10^{-6}			4.2	1.20×10^{-5}
Ni	1.87×10^{-7}			84	1.57×10^{-5}
			$\sum = 1.31 \times 10^{-2}$		$\sum = 28.6 \times 10^{-6}$

WCOB

Metals	CDI (mg kg^{-1}day^{-1})	RfD (mg kg^{-1}day^{-1})	HQ	SF (mg kg^{-1}day)	ELCR
Non carcinogenic metals					
Al	1.50×10^{-5}	1.43×10^{-3}	1.05×10^{-2}		
Cr	5.09×10^{-6}	2.86×10^{-5}	1.78×10^{-1}		
Mn	8.98×10^{-8}	1.43×10^{-5}	6.28×10^{-3}		
Carcinogenic metals					
As	2.23×10^{-8}			15.1	3.37×10^{-7}
Cd	5.38×10^{-9}			6.3	3.39×10^{-8}
Cr	5.09×10^{-6}			4.2	2.14×10^{-5}
Ni	2.04×10^{-7}			84	1.71×10^{-5}
			$\sum = 1.95 \times 10^{-1}$		$\sum = 38.8 \times 10^{-6}$

Table 3. Estimation of human health risk in children due to particulate bound elements from PM$_{2.5}$ emitted from WCOB and ULSD (Betha et al., 2011b)

As shown in Tables 2 and 3, the levels of non-carcinogenic risk (total HQ) were estimated to be 0.06 for ULSD and 0.08 for WCOB and carcinogenic risk (total ELCR) to be 12.3 x 10^{-6} for ULSD and 16.6 x 10^{-6} for WCOB for adults. In the case of children, non-carcinogenic and carcinogenic risks for both the fuels are higher than those in adults. Total HQ was estimated to be 0.13 for ULSD and 0.2 for B100, while total ELCR was 28.6 x 10^{-6} for ULSD and 38.8 x 10^{-6} for WCOB. It implies that 28 to 29 children or 12 to 13 adults in a million can get cancer after exposure to the toxic trace metals in PM$_{2.5}$ emitted from the combustion of ULSD. In the case of biodiesel, it is even higher, 38 to 39 children or 16 to 17 adults out of a million can get cancer after exposure to PM$_{2.5}$ by B100 fuel.

Element	ULSD (mg/m³)	WCOB(mg/m³)
Al	147.6	96.3
Mn	1.4	0.6
Cr	18.3	32.7
Ni	1.2	1.3
Cd	0.06	0.03
As	0.4	0.14

Table 4. Concentration of particulate bound elements in raw exhaust of a stationary engine

From the results it can be deduced that the non-carcinogenic risk indicated by HQ was higher for WCOB compared to ULSD for both groups of people. However, for both ULSD and WCOB, the total HQ was very low for adults compared to children and for both the groups total HQ was below acceptable levels, (Acceptable levels for total HQ =1). On the other hand carcinogenic risk indicated by ELCR was found to be much higher than the acceptable limit for both groups and for both fuels (i.e., 1 in a million) and that ELCR for WCOB was greater than ULSD. From the risk assessment results made in this study, it appears that exposure to $PM_{2.5}$ emitted from biodiesel poses higher risk when compared to $PM_{2.5}$ emitted from ULSD. However, it is to be noted that in this study the carcinogenic risk due to particulate bound elements was used as a measure to evaluate the total carcinogenic risk. A more comprehensive and extensive research needs to be done to evaluate the complete risk assessment including many other carcinogenic compounds such as PAHs and nitro-PAHs. Studies have shown that PAH emissions from biodiesel are very much lower compared to diesel (Jalava et al., 2010; Karavalakis et al., 2011; Lin et al., 2011; Turrio-Baldassarri et al., 2004). Therefore, the total carcinogenic risk of WCOB exhaust particles might be actually lower than ULSD.

In the case of PAHs the the risk assessment for PAHs that are probable and possible human carcinogens were calculated using petency equivalency factor (PEF) relative to BaP and the CDI calculated from Eq (7). Table 5 shows the PAHs with know PEFs (Collins et al., 1998).

PAH	Group	PEF
Benz(a)anthracene, BaA	2A	0.1
Benz(a)pyrene, BaP	2A	1
Benzo(b)fluoranthene, BbF	2B	0.1
Benzo(k)fluoranthene, BkF	2B	0.01
Indeno(1,2,3-cd)pyrene, Ind	2B	0.1
2A: Probable Human Carcinogen 2B: Possible Human Carcinogen		

Table 5. Classification of PAHs by IARC and Potency equivalency factor (PEF)

Carcinogenic risk due to individual PAHs is calculated as product of CDI and PEF. The total carcinogenic risk is the summation of individual risk.

5. Summary

Particle. physical and chemical properties play a key role in determining the health effects associated with PM emissions. Smaller particles can penetrate deep inside the alveolar regions of lungs. Bio-available particulate-bound compounds pose serious health problems. WCOB had lower PNC compared to that of ULSD. However, WCOB had a higher fraction of nucleation mode particles relative to that of ULSD, and therefore, a large fraction of PM emitted from WCOB can deposit in respiratory system compared to DPM. Unlike other types of biodiesel WCOB has higher metal concentrations both in the fuel as well as particulate emissions because of the nature of feedstock. Metals are leached into the oil during cooking and also from cooking utensils. Health risk inhalation of PM was calculated by assessing the CDI estimated using the concentration of particulate-bound compounds and the deposition efficiency of PM in human body, which indicates that WCOB has higher health risk compared to ULSD in terms of particulate bound elements. However, when PAHs are also taken into consideration it can either increase or decrease the relative health risk of WCOB particles depending on the PAHs emission concentrations from both the fuels.

Author details

Raghu Betha, Rajasekhar Balasubramanian and Guenter Engling

Department of Civil and Environmental Engineering, Faculty of Engineering, National University of Singapore, Singapore

References

[1] Abdul-Khalek, I., Kittelson, D.B., Brear, F., 1999. The Influence of Dilution Condition on Diesel Exhaust Particle Size Distribution Measurements. SAE Technical Paper.DOI: 10.4271/1999-01-1142

[2] Alfuso S., Auriemma, M., Police, G., Prati M. V., 1993. The effect of methyl ester of rapeseed oil on combustion and emissions of DI engines, SAE Paper, 932801.

[3] ASTM-D-975, 2006. Standard specification for diesel fuel oils, ASTM.

[4] Atadashi, I.M., Aroua, M.K., Abdul, A.A., 2010. High quality biodiesel and its diesel engine application: A review. Renewable and Sustainable Energy Reviews, 14, 1999-2008.

[5] Atta, M.B., Elsebaie, E.A., Noaman, M.A., Kassab, H.E., 1997. The effect of cooking on the content of heavy metals in fish (Tilapia nilotica). Food Chemistry 58, 1-4.

[6] Bagley, S.T., Gratz, L.D., Johnson, J.H., McDonald, J.F., 1998. Effects of an oxidation catalytic converter and a biodiesel fuel on the chemical, mutagenic,andparticle size

characteristics of emissions from a diesel engine. Environmental Science and Technology, 32, 1183–1191.

[7] Betha, R., Balasubramanian, R., 2011a. A study of particulate emissions from a stationary engine fuelled with ultra-low sulfur diesel blended with waste cooking oil-derived biodiesel. Journal of Air and Waste management association, 61, 1063-1069.

[8] Betha R., Balasubramanian R., 2011b. Particulate emissions from stationary engine: Characterization and risk assessment. Atmospheric Environment. 45, 5273 – 5281.

[9] Boehman, L.A., Morris, D., Szybist, J., 2004. The impact of bulk modulus of diesel fuels on fuel injection timing. Energy Fuels, 18, 1877-1882.

[10] Bünger, J., Krahl, J., Weigel, A., Schröder, O., Brüning, T., Müller, M., Hallier, E., and Westphal, G., 2006. Influence of fuel properties, nitrogen oxides, and exhaust treatment by an oxidation catalytic converter on the mutagenicity of diesel engine emissions. Archives of Toxicology, 80, 540-546.

[11] Burtscher, H., 2004, Physical characterization of particulate emissions from diesel engines: a review. Journal of Aerosol Science, 36, 896-932.

[12] Camden Council (Australia). Camdem Council Biodiesel Truck Trial 2005; Final report. Available online: www.camden.nsw.gov.au/files/camden_council_biodiesel_final_report_march2005a.pdfS

[13] Canakci, M., Van Gerpen, J.H., 2001. Comparison of engine performance and emissions for petroleum diesel fuel, yellow grease biodiesel, and soybean oil biodiesel. ASAE Annual international meeting, Sacramento, California, July 28-31, 46, 937-944.

[14] Cardone, M., Prati, M.V., Rocco, V., Seggiani, M., Senatore, A., and Vitolo, S., 2002. Brassica carinata as an alternative oil crop for the production of biodiesel in Italy: Engine performance and regulated and unregulated exhaust emissions. Environmental Science and Technology, 36, 4656-4662.

[15] Carraretto, C., Macor, A., Mirandola, A., Stoppato, A., Tonon, S., 2004. Biodiesel as alternative fuel: experimental analysis and energetic evaluations. Energy, 29, 2195-2211.

[16] Cass, G.R., 1998. Organic molecular tracers for particulate air pollution sources. Trac-Trends in Analytical Chemistry, 6, 356-365.

[17] CFR, 2006. Title 40 Chapter I, part 60 of the Code of Federal Regulations (CFR).

[18] Chung, A., Lall, A.A. Paulson, S.E., 2008. Particulate emissions by a small non-road diesel engine: Biodiesel and diesel characterization and mass measurements using the extended idealized aggregates theory. Atmospheric Environment, 42, 2129-2140.

[19] Collins, J.F., Brown, J.P., Alexeeff, G.V., Salmon, A.G, 1998. Potency equivalency factors for some polycyclic aromatic hydrocarbons and polycyclic aromatic hydrocarbon derivatives. Regulatory Toxicology and Pharmacology, 28(1), 45-54.

[20] Correa, S.M., Arbilla, G.., 2006. Aromatichydrocarbonsemissions in diesel and biodiesel exhaust. Atmospheric Environment, 40, 6821–6826.

[21] CRC, 2002. CRC PM Workshop Summary. CRC Workshop on Vehicle Exhaust Particle Emissions Measurement Methodology, Manchester Grand Hyatt Hotel, San Diego, California.

[22] Demirbas, A., 2008. Biodiesel: a realistic fuel alternative for diesel engines. London, United Kingdom: Springer

[23] Di, Y., Cheung, C.S., Huang, Z., 2008. Experimental investigation on regulated and unregulated emissions of a diesel engine fueled with ultra-low sulfur diesel fuel blended with biodiesel from waste cooking oil. Science of the Total Environment, 407, 835-846.

[24] Di, Y., Cheung, C.S., Huang, Z. 2009a. Experimental investigation of particulate emissions from a diesel engine fueled with ultralow-sulfur diesel fuel blended with diglyme. Atmospheric Environment, 44, 55-63.

[25] Di , Y., Cheung, C.S., Huang, Z. 2009b. Comparison of the Effect of Biodiesel-Diesel and Ethanol-Diesel on the Particulate Emissions of a Direct Injection Diesel Engine. Aerosol Science and Technology, 43, 455-465.

[26] Di, Y., Cheung, C.S., Huang, Z.H., 2010. Experimental investigation of particulate emissions from a diesel engine fueled with ultralow-sulfur diesel fuel blended with diglyme. Atmospheric Environment, 44, 55-63.

[27] Donaldson, K., Stone, V., Gilmour, P.V., Brown, D.M., MacNee, W., 2000. Ultrafine particles: mechanisms of lung injury. Philosophical Transactions of The Royal Society of London Series A-Mathematical Physical and Engineering Sciences, 58, 2741-2750.

[28] Durbin T., Norbeck J. M., 2000. Effects of biodiesel, biodiesel blends, and a synthetic diesel on emissions from light and heavy-duty diesel vehicles. Environmental Science and Technology, 34, 349-355.

[29] Durbin, T.D., Cocker, D.R., Sawant, A.A., Johnson, K., Miller, J.W., Holden, B.B., Helgeson, N.L., Jack, J.A., 2007. Regulated emissions from biodiesel fuels from on/off-road applications. Atmospheric Environment 41, 5647-5658.

[30] Dwivedi, D., Agarwal, A.K., Sharma, M., 2006. Particulate emission characterization of a biodiesel vs diesel-fuelled compression ignition transport engine: A comparative study. Atmospheric Environment, 40, 5586-5595.

[31] EN-590, 2004. Autmotive fuels - diesel -requirements and test methods, UNE

[32] EPA, 1999. Analysis of the Impacts of Control Programs on Motor Vehicle Toxics Emissions and Exposure in Urban Areas and Nationwide: Volume 1, EPA420-R-99-029, November 1999

[33] EPA, 2002. Health Assessment Document for Diesel Exhaust: Office of Research and Development, Sept.2002.

[34] Fazal, M.A., Haseeb A.S.M.A., Masjuki, H.H., 2011. Biodiesel feasibility study: An evaluation of material compatibility; performance; emission and engine durability. Renewable and Sustainable Energy Reviews,15, 1314-1324.

[35] Ferella, F., Di Celso, G.M., De Michelis, I., Stanisci, V., Veglio, F., 2010. Optimization of the transesterification reaction in biodiesel production. Fuel, 89, 36-42.

[36] Giracol, J., Passarini, K.C., da Silva, S.C., Calarge, F.A., Tambourgi, E.B., Santana, J.C.C., 2011. Reduction In Ecological Cost Through Biofuel Production From Cooking Oils: An Ecological Solution For The City Of Campinas, Brazil. Journal of Cleaner Production, 19, 1324-1329.

[37] Guerrassi, N., Dupraz P., 1998. A common rail injection system for high speed direct injection diesel engines. SAE Technical paper.DOI: 10.4271/980803

[38] Gunvachai K., Hassan, M.G., Shama, G., Hellgardt, K., 2007. A New Solubility Model to Describe Biodiesel Formation Kinetics. Process Safety and Environmental Protection, 85, 383-389.

[39] Hansen, J., Sato,M., Reudy, R., Lacis, A., Oinas, V., 2000. Global warming in the twenty-first century: An alternative scenario." Proceedings of the National Academy of Sciences of the United States of America 97(18): 9875-9880.

[40] Hildemann, L.M., Cass, G.R.,Markowski, J.R., 1989. A Dilution Stack Sampler for Collection of Organic Aerosol Emissions - Design, Characterization and Field-Tests. Aerosol Science and Technology 10(1): 193-204.

[41] Holmen, B. A., Ayala A., 2002. Ultrafine PM emissions from natural gas, oxidation-catalyst diesel, and particle-trap diesel heavy-duty transit buses. Environmental Science & Technology 36(23): 5041-5050.

[42] Hu, S., Polidori, A., Arhami, M., Shafer, M.M., Schauer, J.J., Cho, A., Sioutas, C., 2008. Redox activity and chemical speciation of size fractioned PM in the communities of the Los Angeles-Long Beach harbor. Atmospheric Chemistry and Physics, 8, 6439-6451.

[43] Janaun, J., Ellis, N., Perspectives on biodiesel as a sustainable fuel. Renewable and Sustainable Energy Reviews, 14, 1312-1320.

[44] Jung, H., Kittelson, D.B., Zachariah, M.R., 2006. Characteristics of SME biodiesel-fueled diesel particle emissions and the kinetics of oxidation. Environmental Science and Technology, 40, 4949–4955.

[45] Kado, N.Y., Kuzmicky, P.A., 2003. Bioassay analyses of particulate matter from a diesel bus engine using various biodiesel feedstock fuels. National Renewable Energy Laboratory 2003 NREL/SR-510-31463.

[46] Kalligeros, S., Zannikos, F., Stournas, S., Lois, E., Anastopoulos, G., Teas, Ch., Sakellapoulos, F., 2003. An investigation of using biodiesel/marine diesel blends on the performance of a stationary diesel engine. Biomass and Bioenergy, 24, 141-149.

[47] Karavalakis, G., Bakeas, E., Stournas, S., 2010. Influence of Oxidized Biodiesel Blends on Regulated and Unregulated Emissions from a Diesel Passenger Car. Environmental Science and Technology, 44, 5306-5312.

[48] Karavalakis, G., Stournas, S., and Bakeas, E., 2009. Light vehicle regulated and unregulated emissions from different biodiesels. Science of the Total Environment 407, 3338-3346.

[49] Kaufmann, T. G., Kaldor, A., 1999. Catalysis science and technology for cleaner transportation fuels. Workshop on Building the Future of Catalysis, Egmond Zee, Netherlands, Elsevier Science Bv.

[50] Kawashima, L.M., Soares, L.M.V., 2003. Mineral profile of raw and cooked leafy vegetables consumed in Southern Brazil. J. Food Composition and Analysis 16, 605-611.

[51] Kerminen, V.M., Wexler A.S.,1995. The Interdependence of Aerosol Processes and Mixing in Point-Source Plumes. Atmospheric Environment 29(3): 361-375.

[52] Kittelson, D.B., Johnson,J. H., 1991. Variability of Particle Emission Measurements in the Heavy Duty Transient Test. SAE Paper 42: 137-162

[53] Kittelson, D.B., 1998. Engines and nanoparticles: A review. Journal of Aerosol Science. 29, 575-588.

[54] Kittelson D.B., Watts, W.F. Johnson, J.P., 2003. Nanoparticle emissions on Minnesota highways. Atmospheric Environment, 38, 9 – 19.

[55] Knothe, G., Sharp, C.A., Ryan, T.W., 2006. Exhaust emissions of biodiesel, petrodiesel, neat methyl esters, and alkanes in a new technology engine. Energy & Fuels, 20, 403-408.

[56] Korhonen, P., Kulmala, M., Laaksonen, A., Viisanen, Y., McGraw, R., Seinfeld, J.H., 1999. Ternary nucleation of H2SO4, NH3, and H2O in the atmosphere. Journal of Geophysical Research-Atmospheres 104(D21): 26349-26353.

[57] Krahl, J., Munack, A., Bahadir, M., Schumacher, L., Elser, N., 1996. Review: utilization of rapeseed oil, rapeseed oil methyl ester or diesel fuel: exhaust gas emissions and estimation of environmental effects. SAE paper, 962096.

[58] Kuligowski, J., Halperin, K.M., 1992. Stainless-Steel Cookware as a Significant Source of Nickel, Chromium, and Iron. Archives of Environmental Contamination and Toxicology, 23, 211-215.

[59] Kulmala, M.,Laaksonen, A., 1990. Binary Nucleation of Water Sulfuric-Acid System - Comparison of Classical-Theories with Different H2so4 Saturation Vapor-Pressures. Journal of Chemical Physics 93(1): 696-701.

[60] Lapuerta, M., Rodriguez-Fernandez, J., Agudelo, J.R., 2007. Diesel particulate emissions from used cooking oil biodiesel. Bioresource Technology, 99, 731-740.

[61] Lapuerta, M., Armas, O., Rodriguez-Fernandez, J., 2008. Effect of biodiesel fuels on diesel engine emissions. Progress in Energy and Combustion Science 34, 198-223.

[62] Lin, Y. C., Hsu, K.H., and Chen, C.B., 2011a. Experimental investigation of the performance and emissions of a heavy-duty diesel engine fueled with waste cooking oil biodiesel/ultra-low sulfur diesel blends. Energy 36, 241-248.

[63] Lombardi-Boccia, G., Lanzi, S., Aguzzi, A., 2005. Aspects of meat quality: trace elements and B vitamins in raw and cooked meats. Journal of Food Composition and Analysis 18, 39-46.

[64] Meng, X.M., Chen G.Y., Wang, Y.H., 2008. Biodiesel production from waste cooking oil via alkali catalyst and its engine test. Fuel Processing Technology, 89, 851-857.

[65] Mercer-Blackman, V., Samiei, H., Cheng, K., 2007. Biofuel Demand Pushes Up Food Prices.Available: http://www.imf.org/external/pubs/ft/survey/so/2007/RES1017A.htm

[66] Munack A., Schroder, O., Krahl, J., Bunger, J., 2001. Comparison of relevant gas emiisions from biodiesel and fossil diesel fuel. Agricultural engineering International: the CIGR Journal of Scientific Research and Development; III; Manuscirpt EE 01 001.

[67] NRC, 1983. Risk Assessment in the Federal Government: Managing the Process. National Research Council. Committee on the Institutional Means for Assessment of Risks to Public Health. National Academy Press, Washington, DC.

[68] Oberdorster, G., 2001. Pulmonary effects of inhaled ultrafine particles. International Archives of Occupational and Environmental Health 74(1): 1-8.

[69] Pahl, G., 2008. Biodiesel—Growing a new energy economy, 2 ed. White River Junction, Vermont: Chelsea Green Publishing Company.

[70] Park, C., Kook, S., Bae, C., 2004. Effects of multiple injections in a HSDI engine equipped with common rail injection system. SAE Technical paper.DOI: 10.4271/2004-01-0127

[71] Pischinger, F.F., 1996. The diesel engine for cars - Is there a future? Fall Technical Conference of the ASME Internal Combustion Engine Division, Springfield, Ohio, Asme-AmerSoc Mechanical Eng.

[72] See, S.W., Balasubramanian, R., 2006. Risk assessment of exposure to indoor aerosols associated with chinese cooking. Environment Research, 102, 197-204.

[73] Song J.T., Zhang, C.H., 2008. An experimental study on the performance and exhaust emissions of a diesel engine fuelled with soybean oil methyl ester. IMechE, 222, 2487-2495.

[74] Stamatelos, A. M., 1997. A review of the effect of particulate traps on the efficiency of vehicle diesel engines. Energy Conversion and Management 38(1): 83-99.

[75] Su, D.S., Muller, J.O., Jentoff, R.E., Rothe, D., Jacob, E., Schlogl, R., 2004. Fullerene-like soot from EuroIV diesel engine: consequences for catalytic automotive pollution control. Topics in Catalysis, 30/31, 241-245.

[76] Sydbom, A., Blomberg, A.,Parnia, S., Stenfors, N., Sandstorm, T., Dahlen, S.E., 2001. Health effects of diesel exhaust emissions. European Respiratory Journal 17(4): 733-746.

[77] Szulczyk K.R., McCarl, B. A., 2010. Market penetration of biodiesel. Renewable and Sustainable Energy Reviews, 14, 2426-2433.

[78] Turrio-Baldassarri, L., Battistelli, C. L., Conti, L., Crebelli, R., De Berardis, B., Iamiceli, A.L., Gambino, M., Iannaccone, S., 2004. Emission comparison of urban bus engine fuelled with diesel oil and biodiesel blend. Science Total Environment, 327, 147-162.

[79] United States Department of Energy (US DOE), 1999. Guidance for Conducting Risk Assessments and Related Risk Activities for the DOE-ORO Environmental Management Program. BJC/OR-271.

[80] Verma, V., Shafer, M. M., Schauer, J. J., Sioutas, C., 2010. Contribution of transition metals in the reactive oxygen species activity of PM emissions from retrofitted heavy-duty vehicles. Atmospheric Environment 44, 5165-5173.

[81] Volckens, J., Leith, D., 2003. Partitioning theory for respiratory deposition of semivolatile aerosols. Annals of Occupational Hygiene 47, 157-164.

[82] Ward, D.S., Eidhammer, T., Cotton, W.R., Kreidenweis. 2010. The role of the particle size distribution in assessing aerosol composition effects on simulated droplet activation. Atmospheric chemistry and Physics, 10, 5435-5447.

[83] Xue, J., Grift, T.E., Hansen, A.C., 2011. Effect of biodiesel on engine performances and emissions. Renewable and Sustainable Energy Reviews, 15, 1098-1116.

[84] Zhang, K. M.,Wexler,A. S., 2002. A hypothesis for growth of fresh atmospheric nuclei. Journal of Geophysical Research-Atmospheres 107(D21).

[85] Zhang, K.M., Wexler,A. S., 2004. Evolution of particle number distribution near roadways - Part I: analysis of aerosol dynamics and its implications for engine emission measurement. Atmospheric Environment 38(38): 6643-6653.

[86] Zhu, L., Zhang, W.G., Liu, W., Huang, Z., 2010. Experimental study on particulate and NOx emissions of a diesel engine fueled with ultra low sulfur diesel, RME-diesel blends and PME-diesel blends. Science of the Total Environment 408, 1050-1058.

[87] Zou, L., Atkinson, S., 2008. Characterizing vehicle emissions from burning of biodiesel made from vegetable oil. Environmental Technology, 24, 1253-1260.

Permissions

The contributors of this book come from diverse backgrounds, making this book a truly international effort. This book will bring forth new frontiers with its revolutionizing research information and detailed analysis of the nascent developments around the world.

We would like to thank Prof. Dr. Zhen Fang, for lending his expertise to make the book truly unique. He has played a crucial role in the development of this book. Without his invaluable contribution this book wouldn't have been possible. He has made vital efforts to compile up to date information on the varied aspects of this subject to make this book a valuable addition to the collection of many professionals and students.

This book was conceptualized with the vision of imparting up-to-date information and advanced data in this field. To ensure the same, a matchless editorial board was set up. Every individual on the board went through rigorous rounds of assessment to prove their worth. After which they invested a large part of their time researching and compiling the most relevant data for our readers. Conferences and sessions were held from time to time between the editorial board and the contributing authors to present the data in the most comprehensible form. The editorial team has worked tirelessly to provide valuable and valid information to help people across the globe.

Every chapter published in this book has been scrutinized by our experts. Their significance has been extensively debated. The topics covered herein carry significant findings which will fuel the growth of the discipline. They may even be implemented as practical applications or may be referred to as a beginning point for another development. Chapters in this book were first published by InTech; hereby published with permission under the Creative Commons Attribution License or equivalent.

The editorial board has been involved in producing this book since its inception. They have spent rigorous hours researching and exploring the diverse topics which have resulted in the successful publishing of this book. They have passed on their knowledge of decades through this book. To expedite this challenging task, the publisher supported the team at every step. A small team of assistant editors was also appointed to further simplify the editing procedure and attain best results for the readers.

Our editorial team has been hand-picked from every corner of the world. Their multi-ethnicity adds dynamic inputs to the discussions which result in innovative outcomes. These outcomes are then further discussed with the researchers and contributors who give their valuable feedback and opinion regarding the same. The feedback is then collaborated with the researches and they are edited in a comprehensive manner to aid the understanding of the subject.

Apart from the editorial board, the designing team has also invested a significant amount of their time in understanding the subject and creating the most relevant covers. They scrutinized every image to scout for the most suitable representation of the subject and create an appropriate cover for the book.

The publishing team has been involved in this book since its early stages. They were actively engaged in every process, be it collecting the data, connecting with the contributors or procuring relevant information. The team has been an ardent support to the editorial, designing and production team. Their endless efforts to recruit the best for this project, has resulted in the accomplishment of this book. They are a veteran in the field of academics and their pool of knowledge is as vast as their experience in printing. Their expertise and guidance has proved useful at every step. Their uncompromising quality standards have made this book an exceptional effort. Their encouragement from time to time has been an inspiration for everyone.

The publisher and the editorial board hope that this book will prove to be a valuable piece of knowledge for researchers, students, practitioners and scholars across the globe.

List of Contributors

Joost G. van Bennekom and Hero J. Heeres
University of Groningen, Green Chemical Reaction Engineering, Groningen, The Netherlands

Robertus H. Venderbosch
BTG, Biomass Technology Group, Enschede, The Netherlands

Hayato Tokumoto and Hiroshi Bandow
Department of Chemical Engineering, Osaka Prefecture University, Gakuen-cho, Sakai, Osaka, Japan

Kensuke Kurahashi
Osaka Prefecture University College of Technology, Saiwai, Neyagawa, Osaka, Japan

Takahiko Wakamatsu
Energy and Environment Business Div, Energy Business and Engineering Dept, Osaka Gas Engineering Co., Ltd., Japan

Keysson Vieira Fernandes and Olga Lima Tavares Machado
Universidade Norte Fluminense – Darcy Ribeiro (UENF), Brazil

Maria Catarina Megumi Kasuya, José Maria Rodrigues da Luz and Juliana Soares da Silva
Department of Microbiology, Federal University of Viçosa, Viçosa, Minas Gerais, Brazil

Lisa Presley da Silva Pereira, Hilário Cuquetto Montavani and Marcelo Teixeira Rodrigues
Department of Animal Science, Federal University of Viçosa, Viçosa, Minas Gerais, Brazil

Yanfei Li and Hongming Xu
School of Mechanical Engineering, University of Birmingham, UK
State Key Laboratory of Automotive Safety and Energy, Tsinghua University, China

Guohong Tian
Sir Joseph Swan Centre for Energy Research, Newcastle University, UK

Artur Malinowski, Joanna Czarnocka and Krzysztof Biernat
Automotive Industry Institute, Department for Fuels and Renewable Energy, Warsaw, Poland

Andrey Marchenko, Alexandr Osetrov, Oleg Linkov and Dmitry Samoilenko
National Technical University "Kharkiv Polytechnic Institute", Ukraine

Raghu Betha, Rajasekhar Balasubramanian and Guenter Engling
Department of Civil and Environmental Engineering, Faculty of Engineering, National University of Singapore, Singapore

Printed in the USA
CPSIA information can be obtained
at www.ICGtesting.com
JSHW011407221024
72173JS00003B/443

9 781632 400802